(709)12.-

About Island Press

Since 1984, the nonprofit organization Island Press has been stimulating, shaping, and communicating ideas that are essential for solving environmental problems worldwide. With more than 1,000 titles in print and some 30 new releases each year, we are the nation's leading publisher on environmental issues. We identify innovative thinkers and emerging trends in the environmental field. We work with world-renowned experts and authors to develop cross-disciplinary solutions to environmental challenges.

Island Press designs and executes educational campaigns in conjunction with our authors to communicate their critical messages in print, in person, and online using the latest technologies, innovative programs, and the media. Our goal is to reach targeted audiences—scientists, policymakers, environmental advocates, urban planners, the media, and concerned citizens—with information that can be used to create the framework for long-term ecological health and human well-being.

Island Press gratefully acknowledges major support from The Bobolink Foundation, Caldera Foundation, The Curtis and Edith Munson Foundation, The Forrest C. and Frances H. Lattner Foundation, The JPB Foundation, The Kresge Foundation, The Summit Charitable Foundation, Inc., and many other generous organizations and individuals.

Generous support for the publication of this book was provided by Deborah Wiley.

The opinions expressed in this book are those of the author(s) and do not necessarily reflect the views of our supporters.

ADAPTING CITIES
TO SEA LEVEL RISE

ADAPTING CITIES TO SEA LEVEL RISE
GREEN AND GRAY STRATEGIES

STEFAN AL

FOREWORD BY EDGAR WESTERHOF

ISLANDPRESS

WASHINGTON | COVELO | LONDON

Copyright © 2018 Stefan Al

All rights reserved under International and Pan-American Copyright Conventions. No part of this book may be reproduced in any form or by any means without permission in writing from the publisher: Island Press, 2000 M St. NW, Suite 650, Washington, DC 20036

Island Press is a trademark of The Center for Resource Economics.

Library of Congress Control Number: 2018951657

All Island Press books are printed on environmentally responsible materials.

Manufactured in Canada
10 9 8 7 6 5 4 3 2 1

Keywords: breakwater, climate change adaptation, delta city, dunes, floating island, flood management, floodwall, green infrastructure, levee, living shoreline, multipurpose dike, place-making, revetment, sea level rise, seawall, stormwater management, surge barrier, wetlands

CONTENTS

Acknowledgments ix
Foreword by Edgar Westerhof xi
Chapter 1: INTRODUCTION. **1**

PART I: CITY STRATEGIES **23**
Chapter 2: ROTTERDAM, SOUTH HOLLAND, THE NETHERLANDS **24**
Chapter 3: NEW YORK CITY, NEW YORK, USA **32**
Chapter 4: NEW ORLEANS, LOUISIANA, USA **38**
Chapter 5: HO CHI MINH CITY, VIETNAM **44**

PART II: LOCAL STRATEGIES **51**
Chapter 6: HARD-PROTECT STRATEGIES **53**
    6.1. Protect + Reoccupy/Reclaim 54
    6.2. Seawall 58
    6.3. Revetment 62
    6.4. Breakwater 66
    6.5. Floodwall 70
    6.6. Dike. 74
    6.7. Multipurpose Dike 78
    6.8. Surge Barrier 82

Chapter 7: SOFT-PROTECT STRATEGIES **87**
    7.1. Living Shoreline 88
    7.2. Dunes and Beach Nourishment 92
    7.3. Floating Island 96

Chapter 8: STORE STRATEGIES **101**
    8.1. Floodable Plain 102
    8.2. Polder 106
    8.3. Floodable Square 110
    8.4. Stormwater Infiltration. 114

Chapter 9: RETREAT STRATEGIES **121**
    9.1 Raised Ground 122
    9.2 Floodproofing 126
    9.3 Strategic Retreat 130

Chapter 10: CONCLUSION **137**
Notes 139
Index 141

ACKNOWLEDGMENTS

▶ I thank Island Press executive editor Heather Boyer for publishing this book, and Maureen Gately for managing the design and production. I am greatly indebted to flood risk expert Edgar Westerhof, who was an important influence and generously shared his deep expertise and experiences in flood protection and resilience. I also thank Henk Ovink, ambassador for water and principal of Rebuild by Design, who has been a major source of inspiration. Other influences have been Piet Dircke, Adriaan Geuze, Jesse Keenan, Jeroen Aerts, and Dirk Sijmons. My team of research assistants for this book was led by Jae Min Lee and Nick McClintock. Team members included the very talented Elisabeth Machielse, Ying Luo, Naeem Shahrestani, Cari Krol, Jierui Wei, and Sarah Halle. At the University of Pennsylvania, I am thankful for Jonathan Barnett, Gary Hack, Marilyn Jordan Taylor, Matthijs Bouw, Nick Pevsner, and Richard Weller for sharing their knowledge. In Florida, I am appreciative of Robert Daoust, James F. Murley, Tiffany Troxler, Katherine Hagemann, Samantha Danchuk, Elizabeth Wheaton, and Bruce Mowery. I am grateful to everyone involved in the Counterpart Cities project for the Pearl River delta, including Jonathan Solomon, Dorothy Tang, Jason Carlow, Ivan Valin, Iris Hwang, and Rowan Roderick-Jones. I thank Rebecca Jin, Janneke van Kuijzen, and Vera Al for their love and support. Finally, I thank the designers, engineers, and photographers who allowed their images to be included in this book.

FOREWORD

by Edgar Westerhof, Arcadis Director for Flood Risk and Resiliency, North America

▶ If Superstorm Sandy in 2012 sparked a resurgence of one word throughout the US, that word would be *resilience*. The approximately once-in-one-hundred-year storm that hit the New York and New Jersey region on the evening of October 29 turned out to be a nationwide wake-up call, showcasing the vulnerability of low-lying coastal metropolitan areas. While leading newspapers were holding government leaders a mirror about the possible impact of future extreme weather events triggered by climate change and sea level rise, New York City ignited a strong response.

Following the intense and multi-billion-dollar recovery operation, the Sandy-affected region, initially led by academia, responded by looking for success stories, first at the Netherlands, which is considered a world-class leader in water management through its sophisticated and multilayered flood protection system. New York also looked at New Orleans and the lessons learned following the disastrous flooding caused by Hurricane Katrina in 2005. New Orleans had adopted a massive plan to drastically upgrade its floodwalls, protecting the heavily impacted low-lying communities. In record time of seven years New Orleans had repaired and upgraded its system with long stretches of concrete walls, flood gates, and the largest pump station in the world. The solutions were put to the test during Hurricane Isaac in August 2012 and proved well-suited to withstand hurricane-force winds and storm surge. But although they provided protection, the long concrete walls were also disconnecting communities from the water, leaving children to wonder what was behind those massive walls.

My personal connection to waterfront planning came after completion of my studies in water management and infrastructure planning in the Netherlands in the late '90s. My move to New York in 2012 truly propelled my interest in the interplay between land and water. My first impression was one of disbelief. I did not understand how a world-class, high-end coastal metropolitan city could be so ill prepared when it came to the protection of its citizens. My biking during the hot summer of 2012 turned out to be an exercise in observing the uniqueness and character of local communities, while contemplating how these waterfronts would change under 3, 6, or more feet of sea level rise. The bike trips that remain a prominent memory are the ones I took to Lower Manhattan hours prior to Sandy, as well as that the following morning. I saw the many effects of ignoring waterfront risks and wondered how the many other cities and municipalities along the Atlantic Coast would fare in such a storm.

From the very beginning, the Sandy-affected region acknowledged it had to follow an approach that was focused on connectivity, multifunctionality, and value, rather than on structures that would disconnect communities from the water. The region acknowledged it had to better understand longer term climate implications and the impact on the existing urban system. New York started a citywide exercise to capture vulnerabilities and interdependencies, looking at critical assets in every borough. The results of this study were reported in the Special Initiative for Rebuilding and Resilience (SIRR), which to date speaks to the far-reaching ambition to make the region less vulnerable to extreme weather events and to make communities more resilient. When completed, the SIRR report provided the agenda of what was about to become a dazzling resilience campaign, addressing many of the neighborhoods and critical functions such as hospitals, transportation systems, and power supply. Several studies focused on the longer term, such as those of Coney Island Creek, Jamaica Bay, and Lower Manhattan's Seaport District, contemplating the interplay with short-term recovery and hardening needs.

The city, state, and federal government, through the Department of Housing and Urban Development (HUD), acknowledged the importance of rethinking planning processes to effectively advance the region's climate ambition. In early 2013 HUD brought the Rebuild by Design competition to the region and with the competition an armada of architects, planners, and engineers who volunteered their time to study and design many resilience cases. Multidisciplinary teams collaborated with stakeholders and communities, trying to align needs and ambition while not paying too much attention to the maze of rules that comes with any intervention, even as small as a bench in a park. For nearly a year, participants were exposed to a bohemian vibe of learning by doing, while blending community-driven, climate-robust, and cost-effective ideas into schematic designs.

The most important lesson as the co-lead of Rebuild by Design for my company, Arcadis, was that the adopted outcomes all were the result of an intense and collaborative design process. Rebuild by Design showed that the process was definitely not a straight line going from A to B, but rather a journey exploring shared objectives and stories that helped build trust with stakeholders and the communities. Now in 2018, all winning proposals are working their way to implementation, with the BIG U project (see chapter 1) aiming to break ground in 2018 in the Lower East Side.

Over the next 15 years, nearly all the world's net population growth is expected to occur in urban areas. By the middle of the century, about 70% of the world's population will live in coastal urban metropolitan areas. The global urbanization trend puts tremendous pressure on existing waterfront infrastructure and assets. Maximizing the capacity and use of urban space and increasing challenges and demands puts pressure on city governments to integrate urban functionality to effectively support growth. In this respect, the New York City region follows the trends of delta and coastal cities worldwide. However, where coastal cities such as Rotterdam have, over time, adopted a complex multilayered suite of flood protection solutions, most US cities are challenged to address the threats of climate change and prioritize and integrate short- and long-term resiliency measures effectively into the existing urban fabric. With all coastal cities, New York is challenged to address the threats of climate change and sea level rise. Urban centers must prioritize and effectively integrate long-term resiliency measures into the existing urban fabric while upgrading functionality to current needs and facilitating immediate needs of businesses to ensure their competitive advantage and increase their overall performance.

Building city resilience is no longer about the protection of urban waterfront edges with hard-engineered structures that aim to divide water from land. Cities worldwide are starting to see the community and economic benefit of adopting a more sustainable systems approach to water management. The paradigm has become to integrate water needs with urbanistic functionalities at multiple scales, while enhancing community needs as a whole. Therefore, cities and stakeholders more often face questions related to the ownership and initiative, pushing the need for interagency and multi-stakeholder collaboration to resolve the well-intended and complicated holistic urban ambitions. It is here that the process principles of collaborative water management, practiced, for example, by water governance authorities in the Netherlands, can help create pathways for urban rejuvenation through community-based public private partnerships. In the next few years resilience planning methodologies will improve to validate solution strategies. Financial arrangements and new government structures will arise to help resolve the massive questions that come with adapting to rising sea levels.

Adapting Cities to Sea Level Rise by Stefan Al captures the essence of the most key water management principles and systematically explains their pros and cons by examining the complex urban interplay. While our behavior with water and water management is deeply embedded in a multitude of global cultures, I believe we all have the obligation to share best practices and explain the processes leading to innovations in order to improve efficiency and, in the end, the quality of the solutions. After reading this book people will better understand that water means collaboration, safety, and prosperity, all at the same time.

CHAPTER 1: INTRODUCTION

▶ Sea level rise is already affecting our cities and, depending on where you live, may soon be coming to your home. In 2012, flooding from Superstorm Sandy devastated the Jersey Shore. In 2017, Hurricane Harvey flooded Houston. Today, in the low-lying areas that surround Miami, even on sunny days, king tides bring fish swimming through the streets. These and other events are typically called *natural* disasters. But overwhelming scientific consensus says they are actually the result of *human*-induced climate change and irresponsible construction inside floodplains.

Aftermath on the Jersey Shore
Flood-damaged homes dot the shore after Superstorm Sandy. Some homes even completely disappeared.
(Source: U.S. Fish and Wildlife Service-Northeast Region, Wikimedia Commons)

Climate change is a severe and growing challenge for twenty-first century cities, from droughts to forest fires and from storms to rising sea levels. The most tangible changes are related to water. Some areas will have too little water; others too much. Coastal cities will be strongly affected by the latter, as rising seas increase the occurrence of disruptive or nuisance flooding in urbanized areas during high tides, not just during storms. Water threats in coastal cities can even come from all directions: increased rainfall from above, rising groundwater from below, floods from rivers, and floods from the sea—all worsened by climate change and land subsidence.

Currently, sea levels are rising on average 3.2 millimeters (0.13 in) per year,[1] but this number is expected to rapidly increase as average temperatures rise, with roughly seven and a half feet of sea level rise expected per degree Celsius of warming. Even the conservative 2 degrees Celsius of temperature increase, the goal of the

New Orleans after Hurricane Katrina
(Source: CAN Europe, Flickr)

Paris Agreement, would likely cause sea levels to rise an average 4.6 meters (15 ft), putting coastal cities at risk and some countries, such as the Maldives, entirely under water. Climate Central estimates that two degrees Celsius of warming would lock in 6.1 meters (20 ft) of sea level rise in Miami, leaving almost the entire city underwater. But a mere 0.91-meter (3 ft) rise in Bangladesh would submerge 20 percent of that country, displacing as many as 30 million people. Rising sea levels alone will usher in an entirely new category of human tragedy: the "climate refugee." Rising water also leads to more powerful storm surges and greater flooding on already vulnerable coastlines, forcing more residents to retreat and relocate.

Sea level rise, even a small increase, can dramatically disrupt the day-to-day operation of cities, potentially threatening people's access to power and safe water. Climate change causes extreme weather events, such as floods, to be more recurring and intense, jeopardizing not only the aboveground transmission and distribution electricity networks, but even areas with buried power lines. Furthermore, power outages can cause wastewater treatment plants to malfunction, leading to potentially unsafe water. But flooding can also threaten the functioning of the water system in other ways. Water and sewage plants are typically located at geographic low points, such as along coastlines, to aid wastewater flowing to plants by gravity, which makes them susceptible to overflows at high tides. In addition, stormwater could flow back into the system's discharge pipes, causing backflow. The higher the water level is, the larger the amount of river water flowing back into the drainage system outfalls. When saltwater breaches a plant's treatment system, it could leave permanent damage. Finally, floods can cause saltwater intrusion in coastal freshwater aquifers, as happened during Hurricane Katrina.[2]

Compounding the problem is the rapid urbanization of coastal areas. Human settlement will continue to favor coastal areas, for their benefits of ports, recreation, fishing, and potentially moderate temperatures. In China, for instance, the fastest growing and largest cities, such as Shanghai, Guangzhou, and Tianjin, all face rising rivers and oceans. These three cities alone have added about 40 million people combined in the last thirty years. With the world urban population expected to grow by another two and a half billion by 2050, the problems will only increase.[3] The World Bank predicts that by 2050, $1 trillion or more of assets will be at risk every year in cities worldwide.[4]

Sea level rise will also threaten natural habitats in coastal areas. As seawater reaches farther inland, it can cause land erosion, flooding of wetlands, saltwater intrusion in aquifers, and contamination of agricultural land, potentially causing the loss of habitat for plants and various species of birds and fish.

The best way to prevent these losses would be to avoid climate change altogether, or to *mitigate* the effects by reducing emission levels of heat-trapping greenhouse gases, or to stabilize them, such as through the efforts of the Paris Agreement. But since we are already realizing the effects of climate change, cities will also need to *adapt* to challenges such as sea level rise. One way to respond to future or present flooding problems is by building flood management infrastructure.

A typical approach to floods deals only with the symptoms of the problem. This approach includes increasing a community's "resilience"—in the shallow definition of the term—by improving the capacity to bounce back from a disaster. Examples include evacuating areas before a disaster, temporarily housing people in emergency shelters, or paying out insurance claims after the damage is done, in order to rebuild afterward. But this is very costly, with billions of dollars of losses from floods. Experts advocate for disaster prevention rather than treatment. The Federal Emergency Management Agency estimates that every dollar spent on the reduction of a community's vulnerability to disasters saves people approximately six dollars in economic losses.[5]

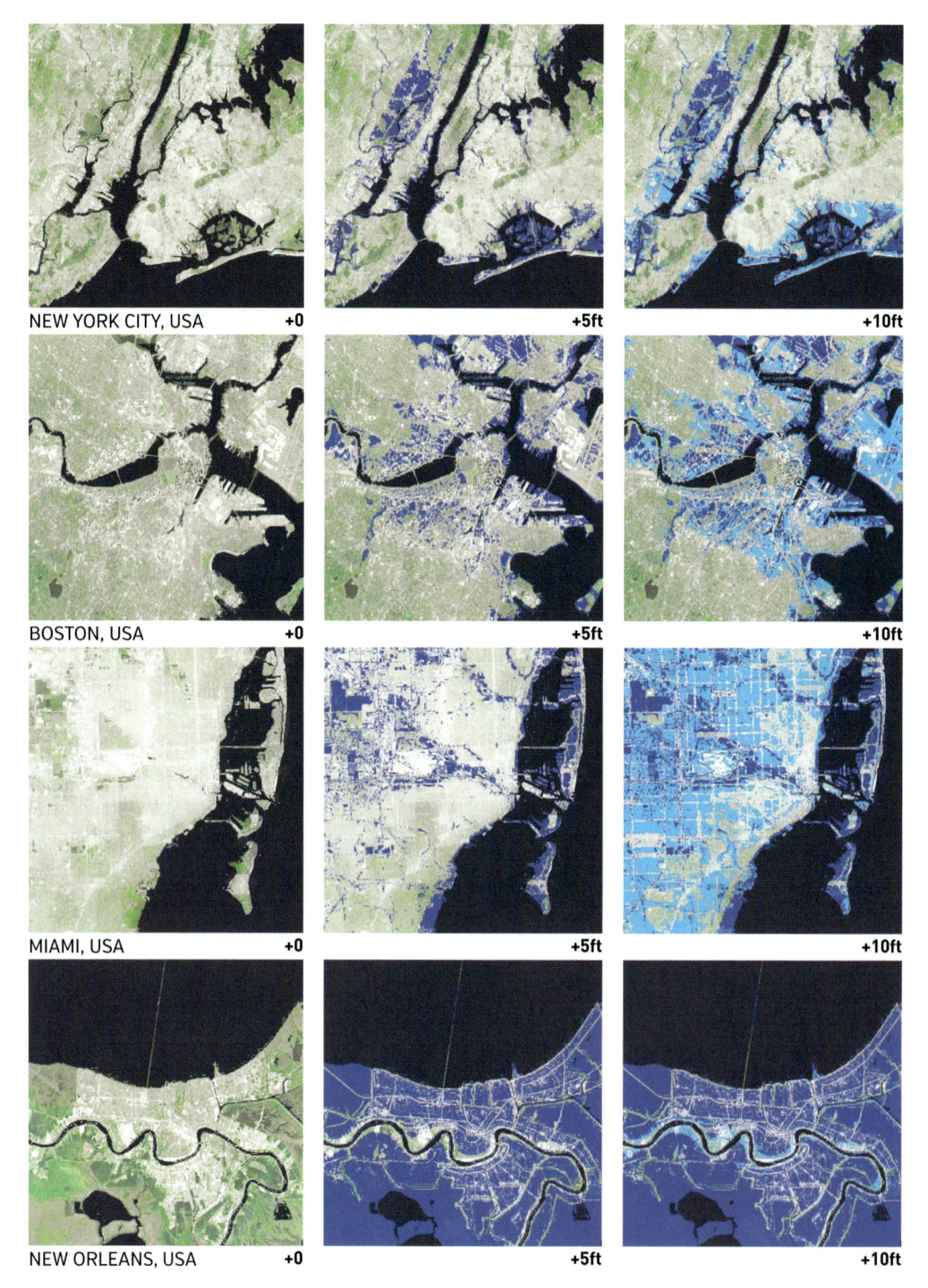

NEW YORK CITY, USA +0	+5ft	+10ft
BOSTON, USA +0	+5ft	+10ft
MIAMI, USA +0	+5ft	+10ft
NEW ORLEANS, USA +0	+5ft	+10ft

WORLD CITIES SEA LEVEL RISE THREAT UNDER MEDIUM- AND HIGH-RISK SCENARIOS

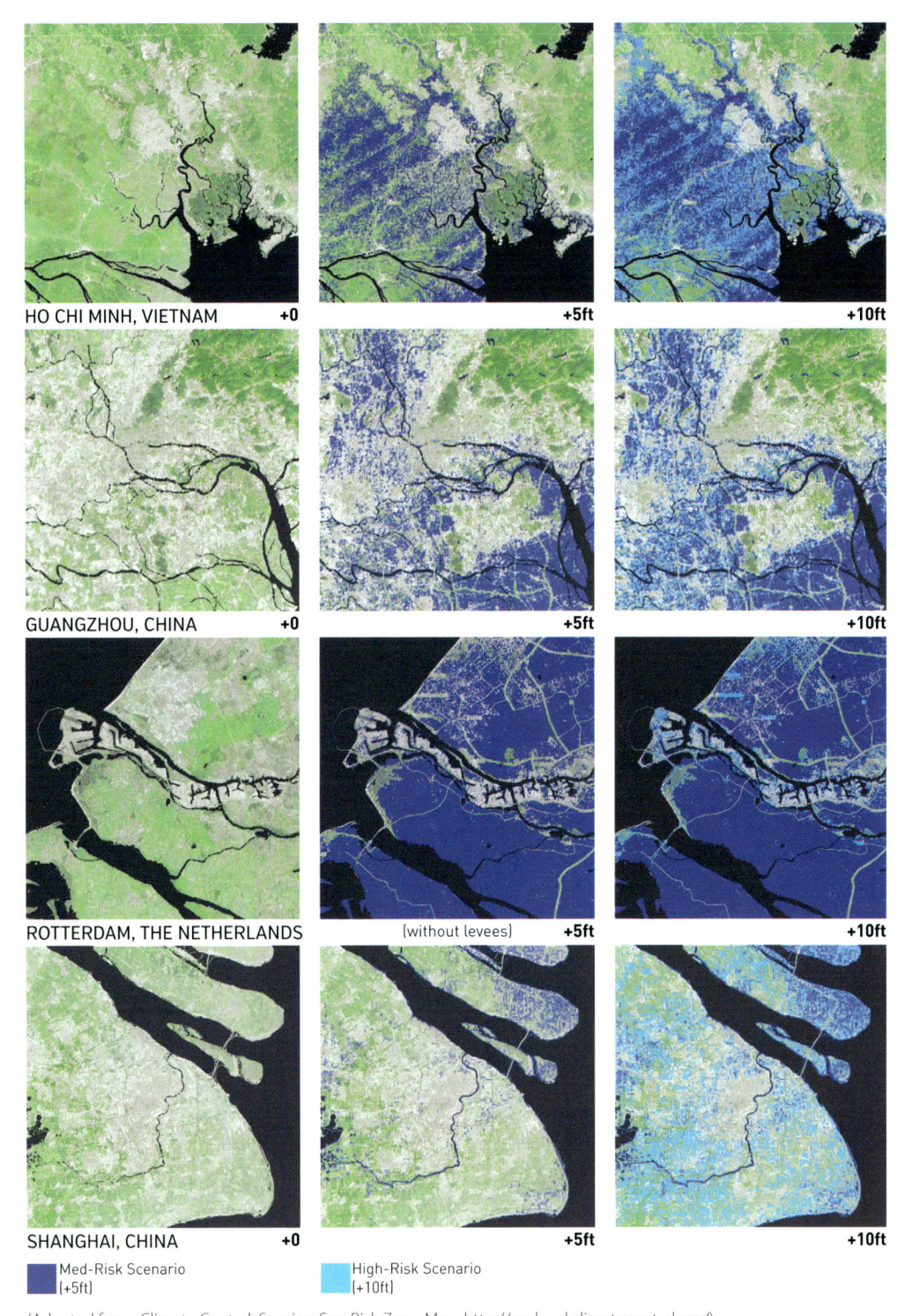

HO CHI MINH, VIETNAM +0	+5ft	+10ft
GUANGZHOU, CHINA +0	+5ft	+10ft
ROTTERDAM, THE NETHERLANDS	(without levees) +5ft	+10ft
SHANGHAI, CHINA +0	+5ft	+10ft

■ Med-Risk Scenario (+5ft)   ■ High-Risk Scenario (+10ft)

(Adapted from: Climate Central, Surging Sea-Risk Zone Map, http://sealevel.climatecentral.org/)

CHAPTER 1: INTRODUCTION | 5

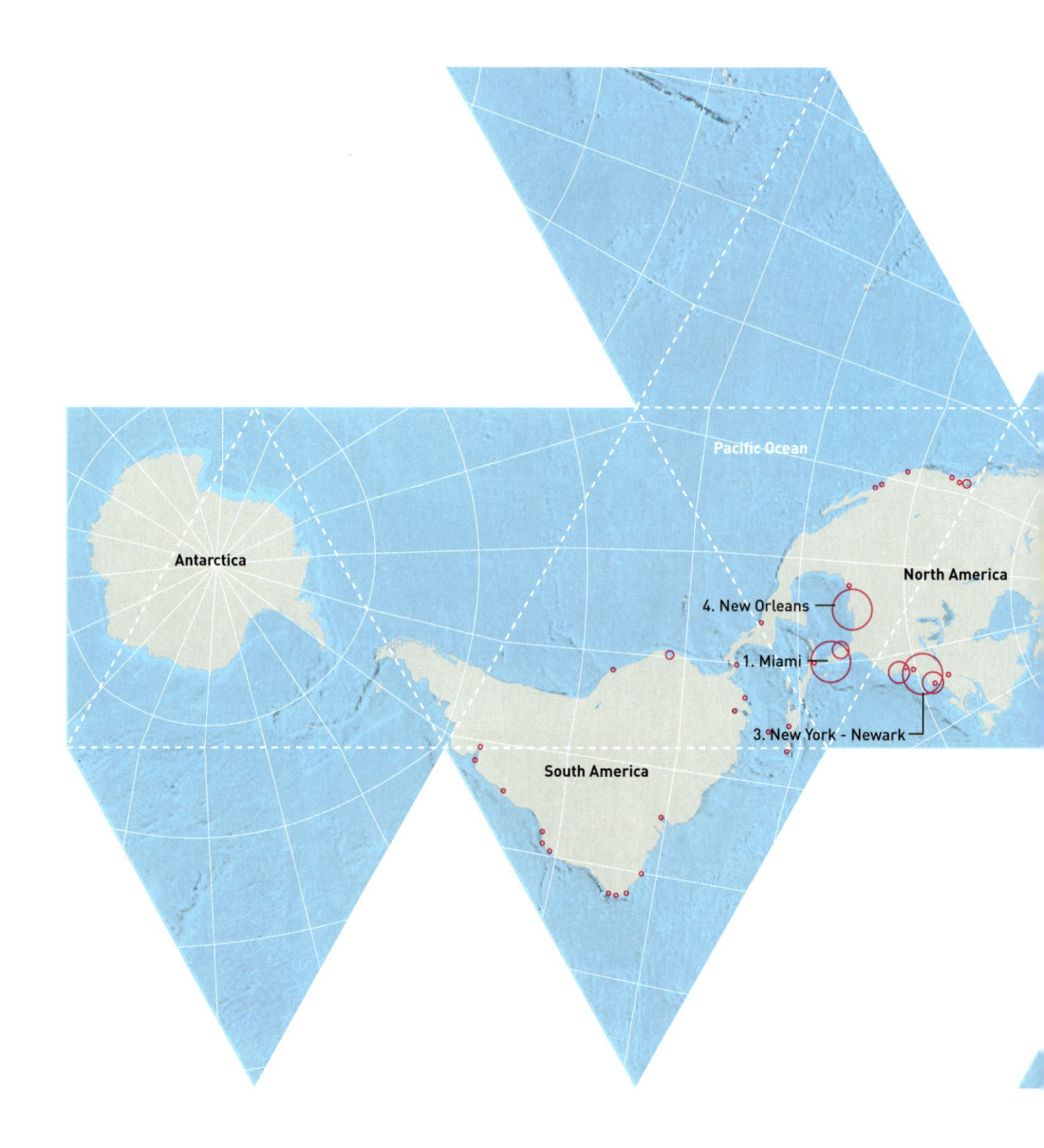

WORLD CITIES SEA LEVEL RISE THREAT
(Adapted from: Stephane Hallegatte et al., 2013, Natural Climate Change, http://ngm.nationalgeographic.com/2015/02/climate-change-economics/coastal-cities-map)

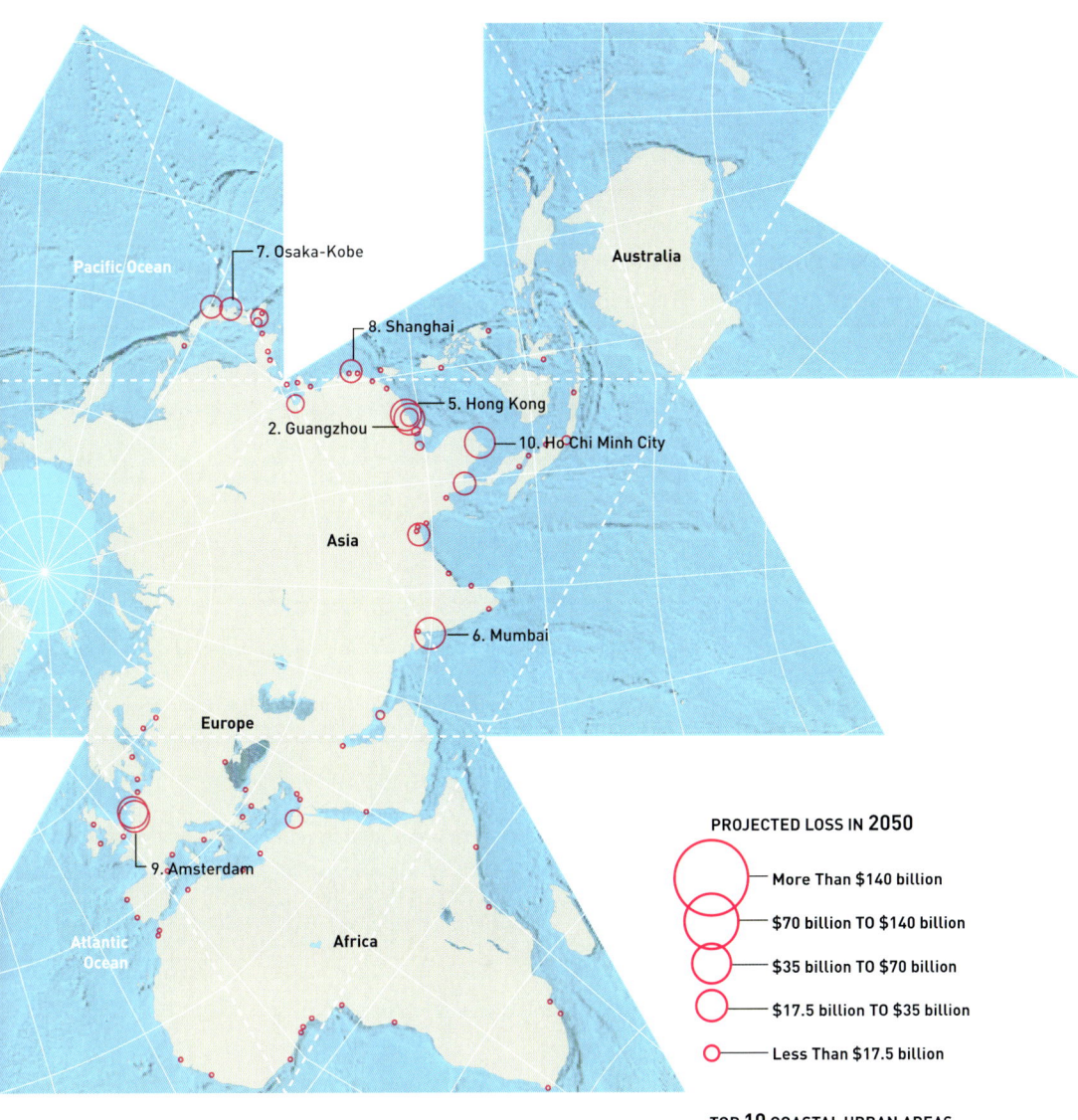

PROJECTED LOSS IN 2050

- More Than $140 billion
- $70 billion TO $140 billion
- $35 billion TO $70 billion
- $17.5 billion TO $35 billion
- Less Than $17.5 billion

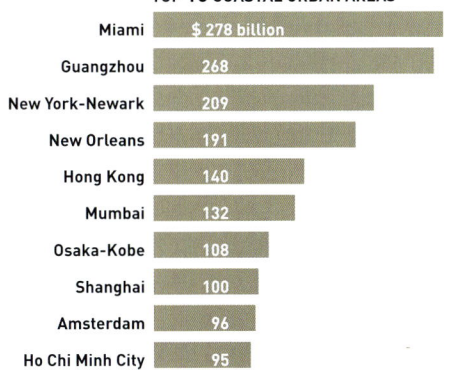

TOP 10 COASTAL URBAN AREAS

City	Projected Loss
Miami	$278 billion
Guangzhou	268
New York-Newark	209
New Orleans	191
Hong Kong	140
Mumbai	132
Osaka-Kobe	108
Shanghai	100
Amsterdam	96
Ho Chi Minh City	95

CHAPTER 1: INTRODUCTION

Another approach to floods—building dikes and seawalls—provides only a short-term stopgap. Imagine a beach house with a beautiful view of the shore. Suddenly, the view is of a concrete wall. A seawall can be a real beach killer in many ways. When a wave strikes a seawall, wave energy is reflected back and carries sand offshore. If the sand is not replenished, the shore will erode and the wall will eventually be undermined, and the environmental harm is done.

Flood management solutions need to be considered beyond their short-term effect to their long-term impact on communities, environments, and economies. They need to respect a community's relationship with the water, as well as the ecological and environmental health of the surrounding area. They should also lead to economic benefits beyond just protecting from floods—for instance, by unlocking the real estate and economic development potential of the newly secured areas, which could help pay

BEACH REPLENISHMENT FAILURE AND BEACH DYNAMICS

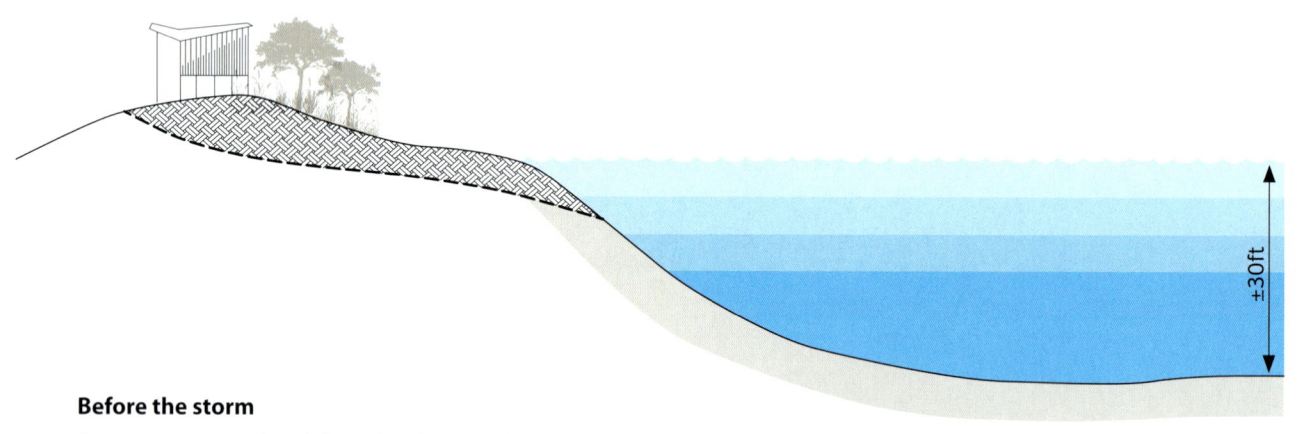

Before the storm

To enjoy sea views, beachfront development has been built on many parts of vulnerable shorelines.

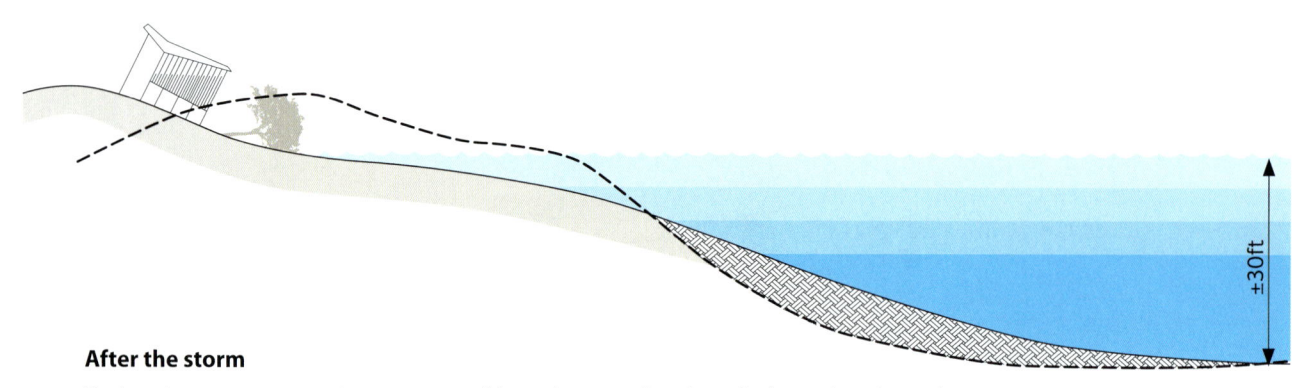

After the storm

During the storm, wave action moves sand from the upper beach to the lower beach, weakening structural integrity underneath beach houses.

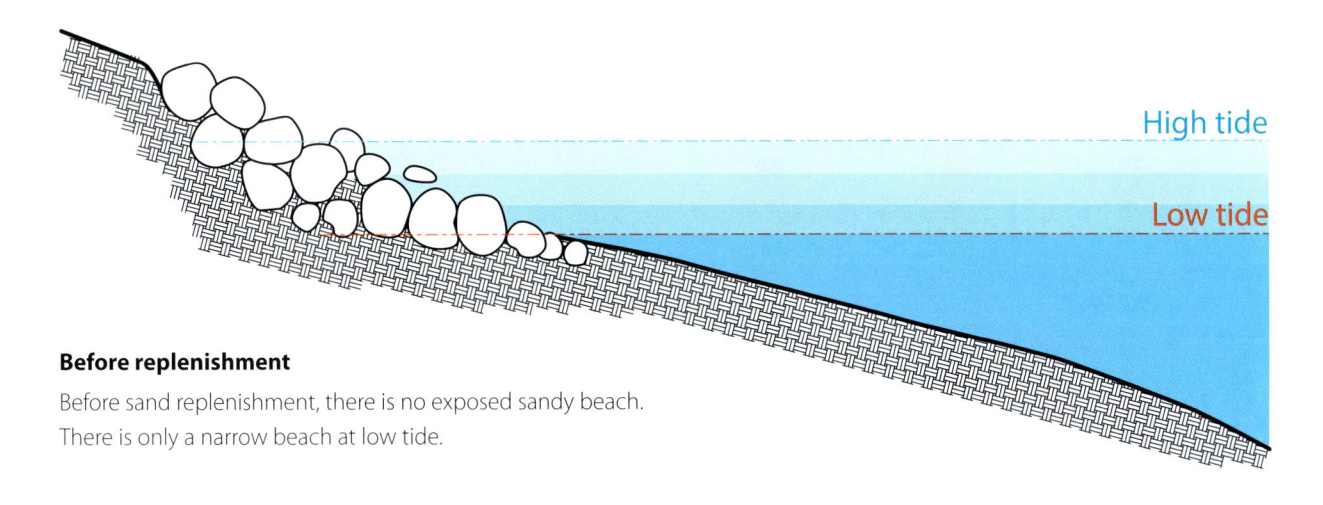

Before replenishment

Before sand replenishment, there is no exposed sandy beach.
There is only a narrow beach at low tide.

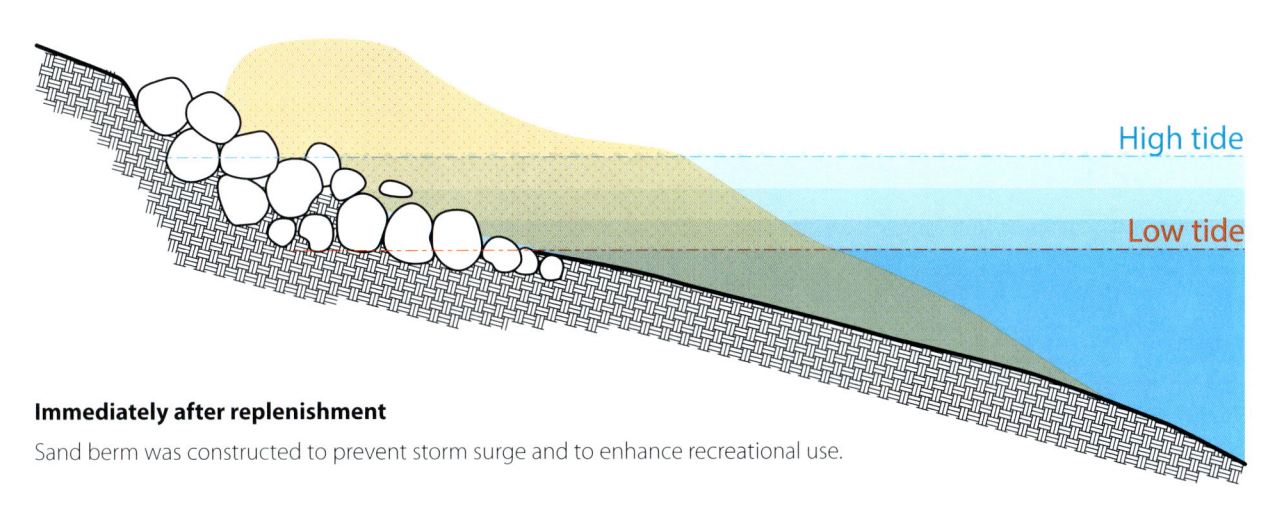

Immediately after replenishment

Sand berm was constructed to prevent storm surge and to enhance recreational use.

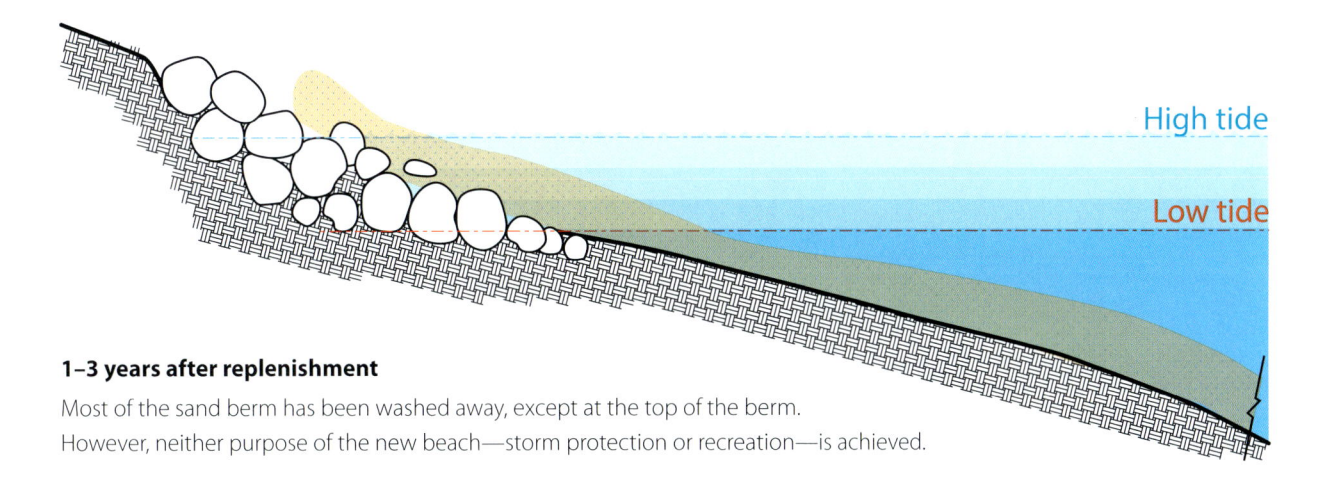

1–3 years after replenishment

Most of the sand berm has been washed away, except at the top of the berm.
However, neither purpose of the new beach—storm protection or recreation—is achieved.

FLOODWALL + BEACH

Before the wall
Sand dunes are often eroded away as a result of natural processes.

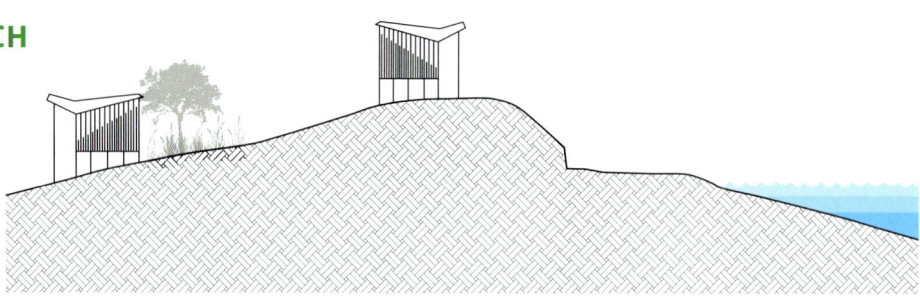

Wall constructed
Homeowners react to the erosion by building a small-scale wall.

2–40 years later
People build higher walls to deal with erosion. In the process the beach disappears.

10–60 years later
Higher-density developments replace cottages and beach houses. A bigger seawall is required for storms.

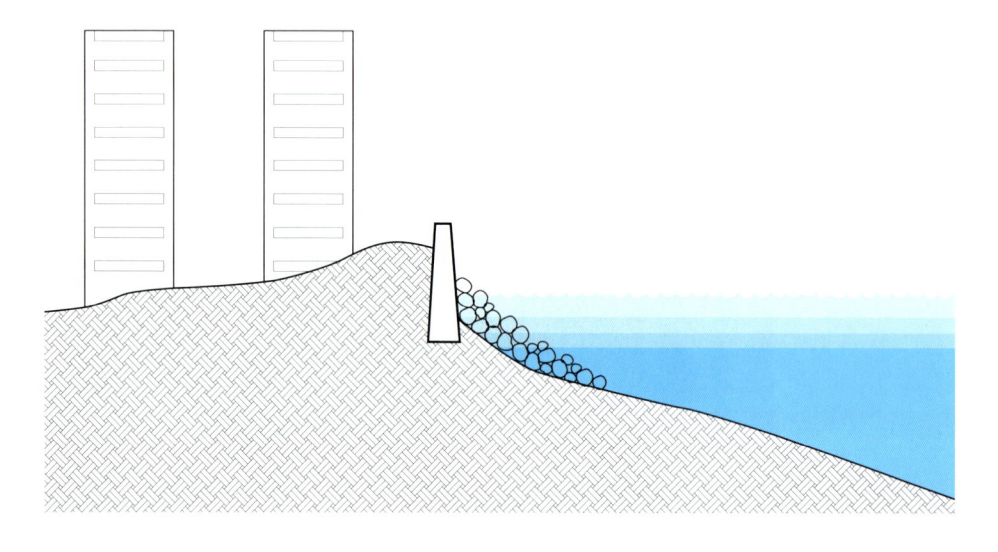

for the infrastructure. In short, flood management infrastructure should not only protect from floods, but should be integrated into the landscape and the public realm. New infrastructure built in or near our cities should be balanced with place-making.

This book, *Adapting Cities to Sea Level Rise,* aims to introduce design responses to sea level rise, drawing from examples around the globe. Going against standard engineering solutions, it argues for approaches that are integrated with the public realm, nature-based, and sensitive to local conditions and the community. In short, it features design responses to building urban resilience that create new civic assets for cities. *Resilience* here is used in the more complex meaning of the term, defined by the Rockefeller Institution's 100 Resilient Cities program as the "capacity of individuals, communities, institutions, business, and systems within a city to survive, adapt, and grow no matter what kinds of chronic stresses and acute shocks they experience."[6] In contrast to the technical definition, this broader definition of resilience has the potential to transform a city.

Fortunately, there are ways in which smart design of flood management infrastructure can make a real difference. For instance, a standard solution of building a revetment—a sloped surface typically built of concrete rocks that dissipate wave energy—has all the charm of a military bunker and can block human access to the shore. In contract to this typical engineering solution, a "designed" solution incorporates other parameters beyond flood protection, such as human use. For instance, the beach town of Cleveleys, United Kingdom, integrated public spaces into flood protection. Instead of a standard revetment, the city built a sinus-shaped structure with amphitheater-like viewing spaces and steps. The steps accentuate the beautiful curvilinear shapes while creating access to the beach and even adding to public space, which is important for a coastal town that relies on tourists.

Nature can play an important role in flood management infrastructure. Instead of reinforced concrete defenses that get the full force of waves and will finally succumb to the undercurrent of the sea, there are solutions that rely on nature's long-term capacity to adapt. Dunes, for instance, are more sustainable sea barriers to absorb the force and velocity of waves, and can also add to the landscape and provide natural habitats. They require only a simple stabilization and wind erosion prevention measure: dune grass, a grass tolerant to high salinity and extreme glare. Landscape architect Ian McHarg was a big fan of this humble plant. "The dune grass, hero of Holland," he wrote in his classic book, *Design with Nature*. Another example is afforestation, planting trees in a drainage basin to help intercept and store water, and thus help reduce a river's discharge and the potential risk of flooding. In addition, coral reefs, seagrass meadows, and mangroves can offer natural ways to prevent coastal erosion, as well as promote biodiversity and water filtration. Today, this philosophy is steaming ahead through the organization EcoShape, which uses "ecosystem services" such as the force of waves and ocean currents to move dredged sand along the southern coast of the Netherlands to replenish beaches and promote dune development—a project called the Sand Engine. It also pioneered the "sandy foreshore," a natural and cheaper alternative to traditional dike reinforcement, in Houtribdijk, the Netherlands. It consists of a large quantity of sand, placed in front of a dike, that reduces the force of waves and also enhances the natural environment and recreational activities.

On the shore, seawalls and revetments can protect the inland areas. But in contrast to these hard edges, living shorelines and dunes can protect the inland area while also creating landscape and promoting biodiversity. Inland, floodable squares help drain floodwater inland, as can flood parks and wetlands, in addition to hosting a range of species, including birds.

This book addresses traditional gray-infrastructure strategies to flood protection, but it also features natural

Sandy foreshore, Houtribdijk

An example of a hybrid gray and green solution, the sandy foreshore uses natural processes to strengthen the existing dike. (Source: EcoShape)

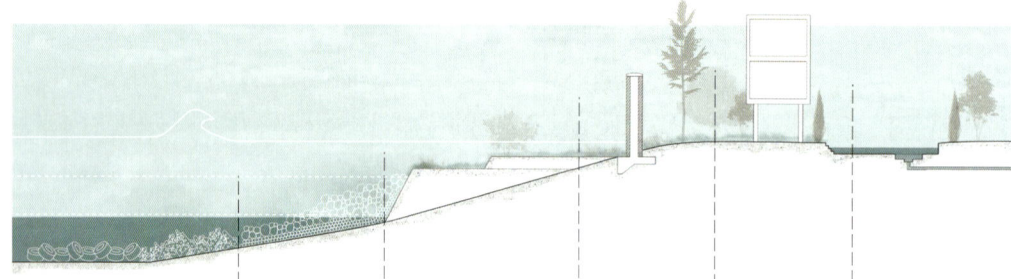

Gray solution

Gray solutions, often developed by civil and environmental engineers, are flood protection structures that are (almost always) permanent. Hard solutions focus on controlling flooding and sea level rise. Examples of hard solutions are seawalls, floodwalls, and revetments. The downside of these projects is their disruption of ecological systems. They are generally expensive and require maintenance.

Traditionally, heavy infrastructural solutions were implemented to minimize coastal flooding. Layers of hard interventions have been implemented by coastal engineers. Concrete breakwaters protect coastline from waves by reducing the intensity of wave action and thereby reducing coastal erosion. Revetments on shorelines reduce the dynamic force of tidal action. Dikes and floodwalls seal off the inland from tsunamis and floods. Existing and new developments can be placed on higher grounds or elevated to prevent further damage to property. Floodable squares and stormwater infiltration beds can help drain floodwater inland.

Green solution

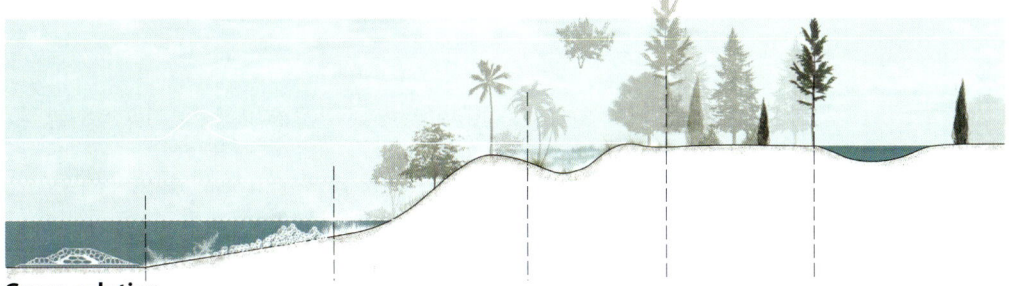

Green solutions utilize ecological and environmental principles and practices to provide flood protection, as well as reduce erosion and stabilize shorelines, while also enhancing habitats and improving aesthetics (as compared to hard solutions). Often, soft solutions are less expensive than hard solutions and lower in maintenance, but they are not permanent and are subject to erosion.

Ecological interventions to flood management depend on natural processes and ecological systems. Natural and artificial breakwaters reduce the tidal action while providing habitats for underwater sea creatures. In contrast to a hard edge such as a seawall, living shorelines and dunes can protect the inland area while maintaining views and access to the waterfront. Inland, flood parks and wetlands help store and release floodwater.

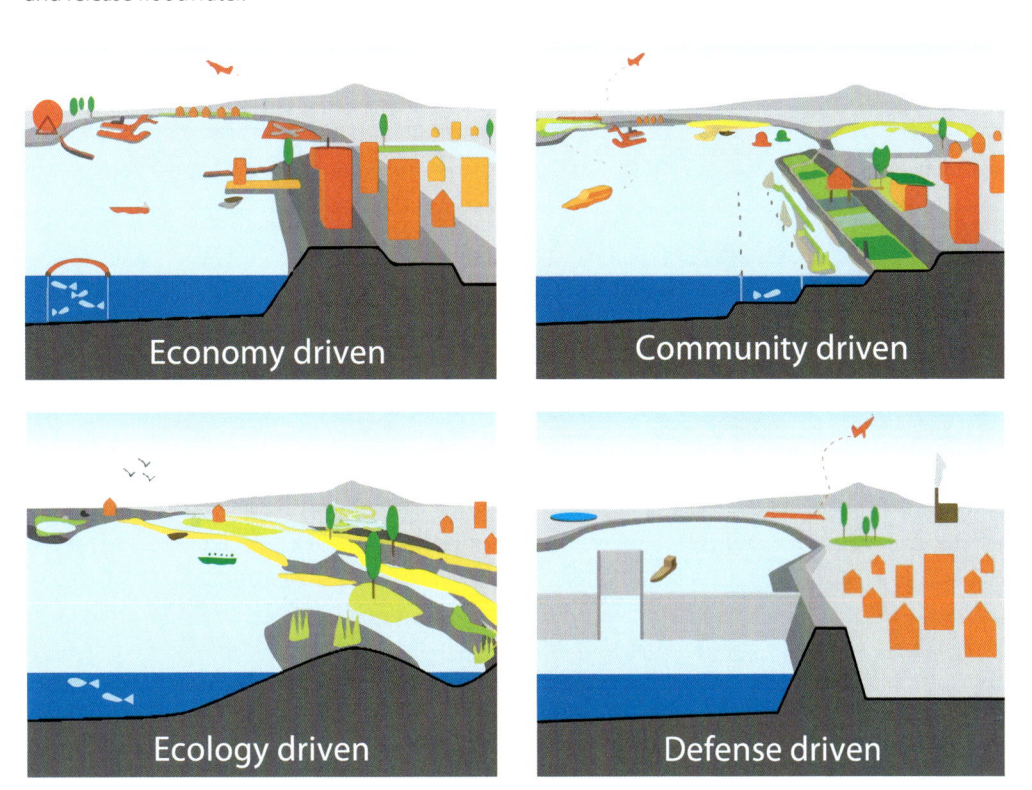

Besides waterfronts driven just by defense, waterfront typologies can also be driven by the economy, community, or ecology.
(Source: Arcadis)

or "green" strategies. All of the functions of gray infrastructure can also be solved by nature. The motto is "soft when possible, hard when necessary."

But, the soft ecological solutions to flood management depend on natural processes and ecological systems, which can take up more space. In addition, they often are vulnerable to human use. Although dune grass is strong in holding together the dunes, it is vulnerable to trampling. Therefore, people should not be allowed to walk on dunes, which makes them a complicated solution in dense, urban areas. Nevertheless, if the dunes are carefully designed with separate walking trails or elevated platforms, they are still viable solutions that not only protect people but also give them recreation space.

In summary, through marrying urban and landscape design with infrastructure, flood management can become a strategic civic asset for cities, communities, and environments. This philosophy considers sea level rise not as a *threat,* but as an *opportunity* to improve our urban areas and landscapes.

FOUR FLOOD MANAGEMENT STRATEGIES:
Hard Protect, Soft Protect, Store, and Retreat

The solutions in this book are organized by four types of flood management strategies that describe their response to dealing with water: 1) hard protect, 2) soft protect, 3) store, and 4) retreat. The first two strategies are gray and green "attack" or "defend" strategies that try to keep the water out, either through holding the line (defending), or by aggressively advancing the line through dredging and land reclamation (attacking). The last two are "accommodation" strategies that let water in, either by storing and controlling floodwater inland, or through "retreat," by either raising ground plains or moving to higher grounds.

The first strategy, hard protect, has traditionally been the one most used in flood risk management, and still is the dominant approach of many organizations, such as the U.S. Army Corps of Engineers. It provides a solid and engineered demarcation between water and land. Examples of hard-engineered protect strategies include seawalls, revetments, breakwaters, floodwalls, dikes (or levees), and surge barriers. To better integrate seawalls into the public realm, they can be designed to be multipurpose, with general seating, recreational activities, and beach access in mind—for instance, as climbing walls or amphitheaters. In addition, other public spaces can be created behind the seawall, such as boardwalks, plazas, parks, or spaces that support commercial activity, helping to create an attractive and investible city. These could make the edge of the seawall an interesting and dynamic experience, rather than something with all the appeal of a bare concrete wall. One such example is Vancouver's seawall, which doubles as a 28-kilometer-long (17.4-mi) greenway—allegedly the world's largest uninterrupted waterfront path—and is one of the city's most popular recreational sites.

The second strategy, soft protect, refers to the use of "soft," or nature-based, systems for flood protection. Examples include living shorelines, dunes, and floating islands. Unlike hard solutions, soft solutions do not have a sharp demarcation line between wet and dry. The in-between area of water and land provides a great habitat for species. Hence, soft-protect strategies can promote biodiversity. But because the soft solutions are less permanent structures, they do not necessarily have the lifespan of hard solutions. Nevertheless, this strategy comes with a natural adaptive capacity we are just now starting to understand.[7]

In addition, the systems strategy is now often combined with hard-protect strategies, making solutions a hybrid gray + green. Soft and hard solutions are not mutually exclusive. They are more frequently used in hybrid solutions and function as a system, working in conjunction to optimize local needs.

One soft-protect case is a living shoreline—a gently sloping natural bank that reduces shoreline erosion, protects coastal ecosystems, and helps reduce storm surge strength along the coastline. Living shorelines use plants and sands and limited hardscape approaches, such as concrete and rocks, to ensure and maintain the natural habitat. They are increasing in popularity over traditional hard-protect strategies, such as bulkheads, because of their added benefits. For instance, the North Carolina Coastal Federation created a living shoreline that not only serves as a buffer for stormwater runoff, but also as a water-purifying lifesaver for the decaying oyster population, whose habitats have been deteriorating.

The third strategy, store, refers to upland water storage. These solutions help solve backflow flooding because of tides and future sea level rise. With the increase of stormwater and flooding upland because of more intense storms due to climate change, this is an important addition to flood protection schemes. The flooding problem is compounded by the impervious surfaces in urban areas, such as asphalt, which do not absorb water. Water storage solutions control water and its movement.

Although belowground storage tanks are typical measures for water storage, aboveground options such as retention and detention basins are viable as well. Aboveground solutions could be integrated with parks and plazas, thus enhancing people's everyday experience with the seasonality and fluctuations of water. These solutions include floodable plains, polders, squares, and stormwater infiltration areas.

For example, floodable squares are lowered urban areas that become pools during heavy rainfall or flooding from the sea or river. They can be used for stormwater storage even in inner cities, such as Benthemplein, which has the first full-scale water square in Rotterdam, the Netherlands. This water reservoir also acts as an urban public space, since the lower areas designed to absorb water can be used for sport or recreation during dry weather.

The fourth strategy, and primary alternative to coastal protection, is retreat—moving away from the risk. Although coastal flood protection generally involves coastal and environmental engineering to manage storm surge and flooding, retreat emphasizes the management of human expectations. Retreat can either be unplanned, as a response to a disaster, or a planned, "strategic" retreat to avoid the costs and/or damage from a disaster. This section will highlight three strategic retreat strategies, including raised grounds, floodproofing, and "de-poldering," a component part of the Room for the River program in the Netherlands.

Raising the ground plane is a strategy that invites water to penetrate waterfront districts, while elevating infrastructure, such as roads, to sustain human use during floods. This technique provides the opportunity for development such as residential, office, hotel, retail, and transit uses. Raised ground planes can support structures and new areas for people to walk, jog, and ride their bikes. Walkways and buildings with varied elevations provide dynamic pedestrian experiences that can work well for office buildings and shopping centers. HafenCity, Germany, an urban regeneration project of a former port area in Hamburg, included the building of new roads and public spaces that are elevated 7.6 m (25 ft) above the normal high tide. It also waterproofed buildings and located new buildings on the raised ground level. Meanwhile, it used the height difference to create viewing decks that provide an interesting experience.

Perhaps the ultimate form of retreat is living on water. Solutions of floating office buildings or neighborhoods already exist. Some buildings are even created as amphibious structures, such as one New Orleans home built by Pritzker Prize-winner Thom Mayne. He designed the base of the house as both a foundation and a hull that provides enough buoyancy to float the entire weight of the house on water.

16 | ADAPTING CITIES TO SEA LEVEL RISE

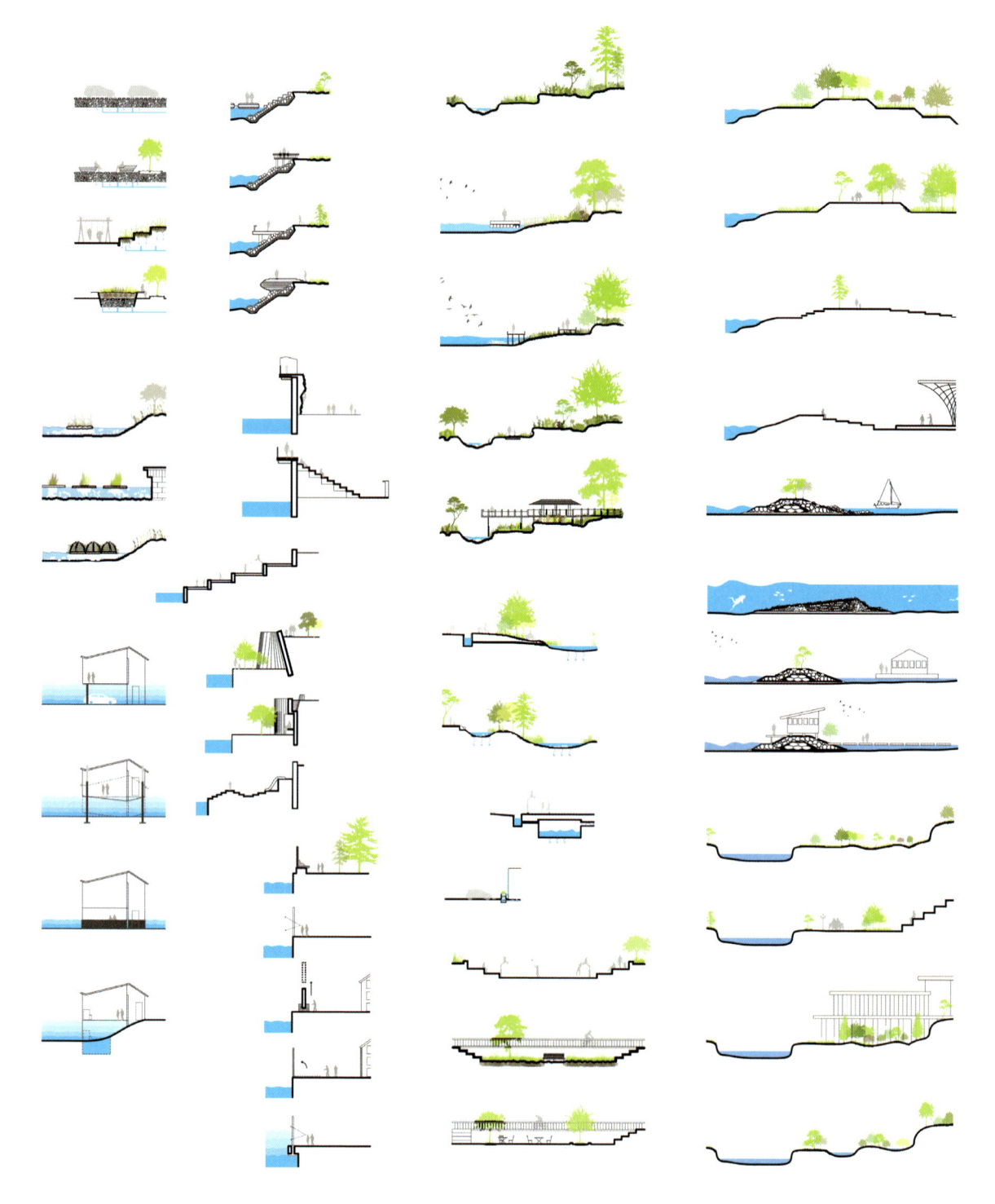

TOOL BOX OF SEA LEVEL RISE INTERVENTIONS

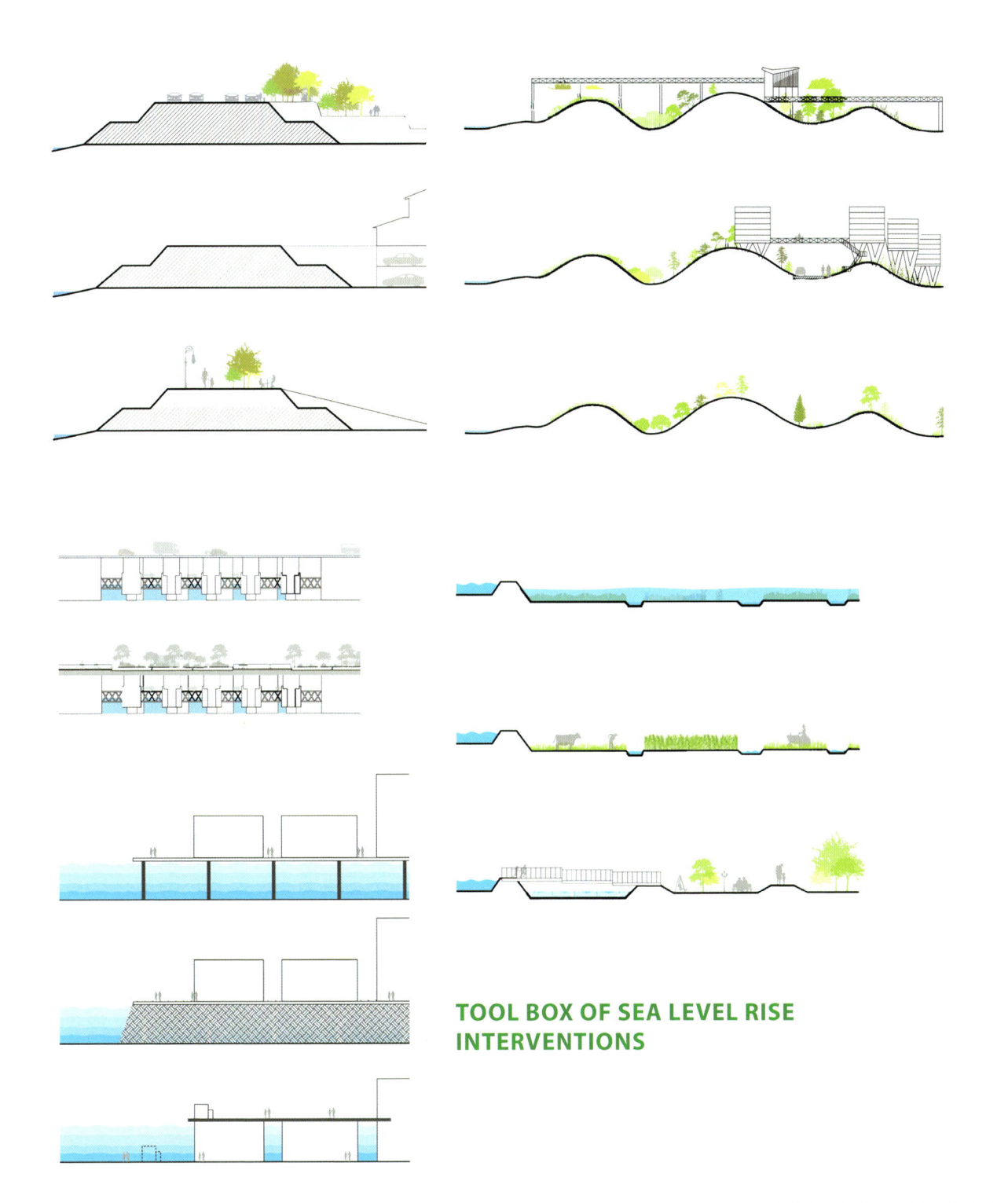

CHAPTER 1: INTRODUCTION | 17

REGIONAL COMPREHENSIVE RESILIENCE PLANS: Rotterdam, New York, New Orleans, and Ho Chi Minh City

The four flood management strategies—hard protect, soft protect, store, and retreat—are not mutually exclusive. They can be combined into comprehensive flood management solutions on both a local and regional level. The first part of the book shows how each of these four fundamental strategies can come together on a citywide scale. Retreat may make sense in urban areas with little capital or few vulnerable populations and urban assets at risk, but it may not be an option in urban areas with high population density and vital infrastructure. Some urban areas may be more at risk of stormwater flooding than others, and hence could focus on store strategies. Hard protect may make sense in a city's densest areas, where soft protect would not be feasible. The lesson is to use the four approaches comprehensively rather than exclusively, and based on a detailed vulnerability and risk analysis. Adaptation of one area may exacerbate flooding elsewhere, and strategies at all scales are related to their existence as part of a watershed, so interventions need to be considered comprehensively. For instance, the national-scale Room for the River project in the Netherlands is connected to local solutions in the city of Rotterdam, such as the water plazas.

Four cities from three continents are featured, each with a robust resilience plan: Rotterdam, New York, New Orleans, and Ho Chi Minh City. These cases are sampled in such a way to show a variety of contexts, including different climatic, urban, and sociopolitical conditions. Every city is different, and lessons need to be considered and strategies applied for the unique conditions of each. For instance, it is highly improbable that the high level of water control achieved by the Rotterdam resilience plan applies to places outside of the Netherlands, a country that has managed its relationship with water for centuries and dedicates more than one percent of its entire gross domestic product to water management. In fact, "water boards" were founded there to manage water as early as the twelfth century, and taxes were raised to protect from floods centuries ago, even before city taxes were levied. This is a unique resource-rich and top-down planning context with an extremely strong control of water that may not apply elsewhere. Some political climates do not favor long-term planning. In other places, there may simply not be the capital. Then there are situations in which a particular charged solution, such as retreat, may be a political impossibility.

Rotterdam is featured first. But the city's flood system is only one of several pillars that protect the city, given the Netherlands's national risk approach. Following the storm surge of 1953, which led to 1,835 fatalities, the Delta Commission was established, tasked to prevent future disasters and reinforce the delta. The commission argued that the costs of protection would be offset by the reduction of flood risk. In order to understand where investments in reinforcements should be made, a scientific and probabilistic approach to flood management infrastructure was implemented. It led to the National Flood Risk Analysis in the Netherlands, called the VNK project.

Essentially, flood risk is a product of the probability of a failure of a flood protection system and the consequences of a failure (flood risk = probability of a failure × consequences of a failure). Mitigation decreases, and urbanization increases, those consequences. Even the most advanced computer models cannot accurately predict when and where a flood defense may fail, given the many factors that are uncertain, such as the precise maximum strength of a structure and the precise load. However, models and statistics can determine a probability of occurrence of failure for all possible combinations of uncertain strengths and uncertain loads. This allows the incorporation of uncertainties into the method.[8]

The second half of the equation, the consequences of a failure, includes running simulation models that calculate various flood scenarios and their flooding characteristics, such as water velocity and water depth, and how flood patterns may interact with landscape features such as railway lines. Combining flood probabilities with consequences gets to flood risk, which in the Netherlands is considered on an economic, individual, and societal level.

The Delta Commission distributes different flood risk levels depending on an area's assets and population. The highest standard for flood protection, at an annual probability of failure of 1/10,000, is in the west of the country—which includes cities such as Amsterdam and The Hague—a region with the highest economic value. The rest of the Netherlands has lower standards, varying from 1/4,000 to 1/1,250 for local watersheds.

Rotterdam is located in an area in which there is an extreme control of water, built to a 1-in-10,000-year flood standard. The city's resilience plan divides the city into districts based on their relative position inside or outside the existing inner-ring dikes and their relative density and age. Newer districts on the periphery of the city will get larger-scale investments in blue/green infrastructure to manage stormwater, elevate particular areas of land, and raise the water table. Older districts with denser building fabric within the older inner-dike system will integrate large-scale building with dike reinforcements, and in some places, the replacement of existing dikes with seawalls and other features that will allow for more development. Older port facilities in the center city, known as the Stadshavens district, will see a combination of dike reinforcement with building integration, and large-scale blue/green infrastructure.

What is unique about Rotterdam is that it is a multi-level, multipurpose, and multi-stakeholder strategy, and has adopted these principles to safeguard both the city's industrial and urban functions—all the while, 90 percent of the city sits below sea level, with some areas 6.1 meters (20 ft) under. The city is a real source of flood management innovation and experimentation, including the first water square and floating pavilion. Even the city's central train station has underground water storage that doubles as a bike parking garage. Finally, there are high levels of collaborative and participatory modes of governance, which include the community in deciding on interventions.

Then comes New York City. Up until Superstorm Sandy in 2012, flood protection was not a primary policy topic, especially considering the trillions of dollars of assets at risk.[9] Following Sandy, which led to the tragic death of forty-three people and the nicknaming of lower Manhattan as SOPO (South of Power), New York City updated its PlaNYC to incorporate initiatives for urban resiliency, including strategies for addressing flooding, storm surge, and sea level rise. Three years later, this plan was updated in OneNYC, which focused on large-scale resiliency. Today, New York has a resilience plan that sets a new standard for U.S. cities, and is a global pioneer in combining community resilience with rebuilding by design. The plan is a great example of "don't let a good disaster go to waste"—yet, it still lacks a regional focus.

However, Dutch flood management expert and economist Jeroen Aerts showed how a plan for a regional barrier to protect from floods—the New York–New Jersey Outer Harbor storm surge barrier—would have been the most cost-effective option to protect New York.[10] But, for political and other reasons, New York abandoned the barrier for a plethora of local strategies, including multi-purpose levees and building codes. As a result, New York is not up to the Rotterdam standard of flood protection. Ultimately, the key resilience question for any city is a political one: exactly how resilient do we want to be?

New York not only chose a different standard of resilience, but also chose a different kind. In 2013, New York City incorporated a view of resilience borrowed from the 100 Resilient Cities program by the Rockefeller Foundation. It not only aims to build resilience to bounce back from shocks such as floods, but also the day-to-day stresses that weaken the fabric of the city, such as an inefficient transit system and chronic food and water shortages. The city became the testing ground for Rebuild by Design, an initiative cosponsored by 100 Resilient Cities that reimagines the process by which communities create solutions to complex and large-scale problems, such as flooding. It takes a collaborative research and design approach—convening stakeholders such as government, business, nonprofit, and community organizations, and led to several detailed design interventions, including the iconic BIG U, a protective system around Lower Manhattan designed by a team spearheaded by Bjarke Ingels Group.

But there may be unintended consequences to government investments in mitigation, as described by researcher Jesse Keenan's new term of "climate gentrification."[11] Although the BIG U, nicknamed the Dry Line, addresses the lack of open space in Manhattan's Lower East Side, it will likely raise property values and may ultimately displace the low-income community. A cautionary tale may be the project that gave it its nickname, the High Line, an elevated park project that was a massive success and helped regenerate the Chelsea district. Nevertheless, critics of the project have pointed out that the benefits of the park have not been experienced evenly across the economic spectrum, because lower-income residents have been displaced by gentrification.[12]

The second city to be featured is also in the United States, but with an entirely different context. More than one-third of New Orleans's land area is wetlands, and a majority of the land area is at or below sea level. The U.S. Army Corps of Engineers, as mandated by the Flood Control Act of 1965, constructed a levee system. Then, in 2005, Hurricane Katrina came. The hurricane caused a storm surge that led to approximately twenty-three breaches in New Orleans's drainage and levee system. Eighty percent of the city was flooded, with some areas 4.6 meters (15 ft) submerged. Nearly 90 percent of the city's residents were evacuated. Nevertheless, hospital officials counted 1,464 deaths, in what was called by some the worst engineering disaster in American history.

New Orleans rebuilt its concrete levees, a classical defense strategy that aims to keep the storm surge out. In addition, it looks to end wetlands loss and expand the city's flood protection systems. Meanwhile, the Resilient New Orleans plan brings together local understanding and global best practices to reduce risk and inequity. New Orleans focuses on the effects of large storms, as well as the daily stresses on the natural and built environment. New Orleans is particularly concerned with sea level rise and coastal flooding as it relates to the issues of poverty and resource allocation within the city, as much of the population at risk for flooding is living in poverty.

Climate gentrification here is an issue as well. After Katrina, approximately 300,000 low-income residents did not return to eir homes, because of expensive rebuilding efforts and increasing flood insurance rates and taxes. Living in high flood-exposure areas such as New Orleans or Miami Beach can become prohibitively expensive. Add to that revitalization efforts that lead to land appreciation, and low-income residents are the first to be displaced.

The fourth resilience plan featured in this book is Ho Chi Minh City's in Vietnam. The city's climate change threats include increased flooding events and saltwater intrusion as a result of sea level rise, temperature increase, and a heat island effect. Compounding the problem is the challenge of rapid urban growth, driven by the migration of rural residents aiming to participate in the city's fast-developing economy, as well as formal and informal settlements built in flood zones. Additionally, there are risks to the economy of Southeast Asia if Ho Chi Minh City is unable to function at its current capacity.

The city used the opportunity of adapting to sea level rise to plan a new dike system. Some of the dike developments in floodplains have been integrated with new urban districts and iconic waterfront development with underground water basins, allowing for future growth that is resilient to floods and storms.

THE FUTURE OF CITY DESIGN IN THE "RISK SOCIETY"

Sociologist Ulrich Beck has argued that the increasing number of industrial accidents, including Chernobyl, as well as the results of human-made climate change, including floods and permanent droughts, are lasting side-effects of our "Risk Society." The projects in this book corroborate the point. Hedging environmental risk has become part of city plans worldwide.

Creating design solutions to improve urban resilience and reduce risk is a rapidly emerging field that exists at the interdisciplinary intersection of engineering, urban design, landscape architecture, architecture, and urban and environmental planning. City planning departments worldwide are adding resilience officers to their teams, who are focused on interdepartmental planning and coordination, and developing "climate-proof" plans resilient to climatic problems ranging from floods to droughts. Universities globally are implementing resilience curricula. Designing flood management infrastructure is already a billion-dollar market. With the impact of climate change increasingly "severe and pervasive,"[13] this industry is bound to grow.

Meanwhile, the Risk Society demands a whole new breed of design. Instead of designing for static situations, designers need to anticipate the uncertainty of the future. Instead of operating in silos, they need to work across disciplines with experts including hydrologists, engineers, and city planners. Rather than working with experts only, they need to collaborate with multiple stakeholders. In short, designers need to coordinate integrated solutions, be part of a collaborative water management process, and design for an uncertain future and risk.

First, designers working in coastal areas need to approach sites by acknowledging that the water comes first. They should first understand the ecological and hydrological aspects of a situation. The urban condition follows, from integrating transportation to urban development to other programmatic demands. Then come the other aspects they can consider in a design, from improving the "image" of a project to improving the social interactions among people following the lessons of Jane Jacobs. This integrated approach includes hydraulic, environmental, economic, social, and aesthetic aspects—all rolled into one.

Secondly, architects need to work collaboratively with stakeholders, including representatives from finance, development, government, insurance, and the community. Not only do collaborative processes bring in local knowledge, they also help buy-in and increase the odds of success of the project. "Collaborative water management" has become a new buzz phrase in the flood management literature.[14] It reflects the most recent approach in the Netherlands, a more holistic response of "integrated water management," and also the philosophy of the Rebuild by Design competition in the United States, which uses the design competition to bring together various teams with multiple stakeholders.

One particular tool aimed to promote collaborative design is the "design charrette," an event in which various disciplines and stakeholders work together on the design of a plan. "Charrette," French for cart, refers to carts that collected the exams of architecture students at the nineteenth-century École des Beaux Arts in Paris. But the term more generally describes the feverish activity the students went through as they rushed to finished drawings and blueprints before the carts wheeled in. But whereas students then worked independently, the point of the modern-day charrette is to sidestep business as usual and to bring people from different disciplinary and political silos together to work collaboratively on developing solutions.

Finally, designers will have to get comfortable with the fact that urban and natural systems are complex and adaptive. They will have to design with uncertainty in mind, think in probabilities and scenarios, and be ready to adapt to change. More than ever, the future is a moving target.

As difficult as all of this may be, this book advocates for designing cities to cope with risk, argues for balancing flood management with place-making, and aims to be an easy-to-use reference for everyone interested in the long-term resilience of their city. Climate change is the new normal, and so are sea level rise, intense storms, extreme heat, forest fires, and other emerging climate-related risks to our society. The book can be read from cover to cover as much as it can be used as a source with information about a specific flood protection strategy or case study. It is meant for architects, engineers, urban designers, landscape architects, policy makers, developers, investors, and communities—in short, for everyone who has a stake in building more-resilient, twenty-first-century cities.

PART I
CITY STRATEGIES

Chapter 2: Rotterdam, The Netherlands

Chapter 3: New York City, New York, USA

Chapter 4: New Orleans, Louisiana, USA

Chapter 5: Ho Chi Minh City, Vietnam

24

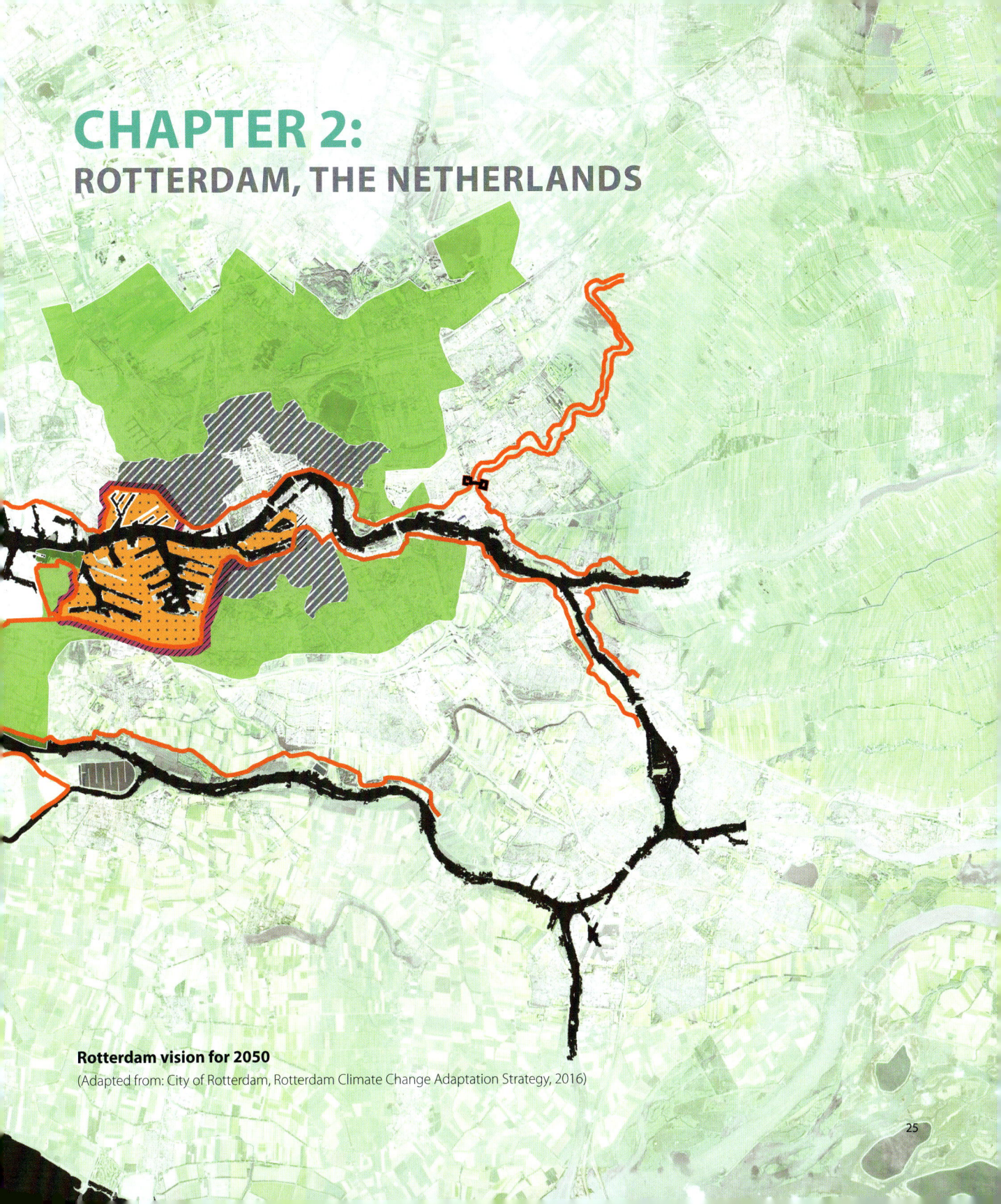

CHAPTER 2:
ROTTERDAM, THE NETHERLANDS

Rotterdam vision for 2050
(Adapted from: City of Rotterdam, Rotterdam Climate Change Adaptation Strategy, 2016)

▶ Ninety percent of Rotterdam is located below sea level, with some areas as much as 6.1 meters (20 ft) below the North Sea. At the same time, the city relies on its functional port, Europe's largest. Rotterdam has overcome its challenges and has become a best practice case for flood resilience.

Rotterdam's resilience plan must be understood within the context of the Dutch Delta Works, an elaborate system of dams, dikes, and barriers, established after the disastrous 1953 flood, when almost two thousand people drowned. This system established a 1/10,000 chance of flooding a year[1] for cities such as Rotterdam and Amsterdam, about 100 times as safe as the average coastal city. One highlight of the Delta Works is the Maeslant Barrier, one of the world's largest moving structures. It closes if Rotterdam is threatened by storm surges but stays open to allow passing ships and the functioning of the port, the lifeblood of the city.

In addition, Rotterdam also benefits from large water reservoirs outside the city and the Room for the River program, a 2006–2015 nationwide project that relocated dykes and reduced the height of groins to create additional space for water within the river's floodplains. Extreme flooding in the future can now be absorbed within the rivers, instead of in the city.

Moreover, Rotterdam's city-level flood protection

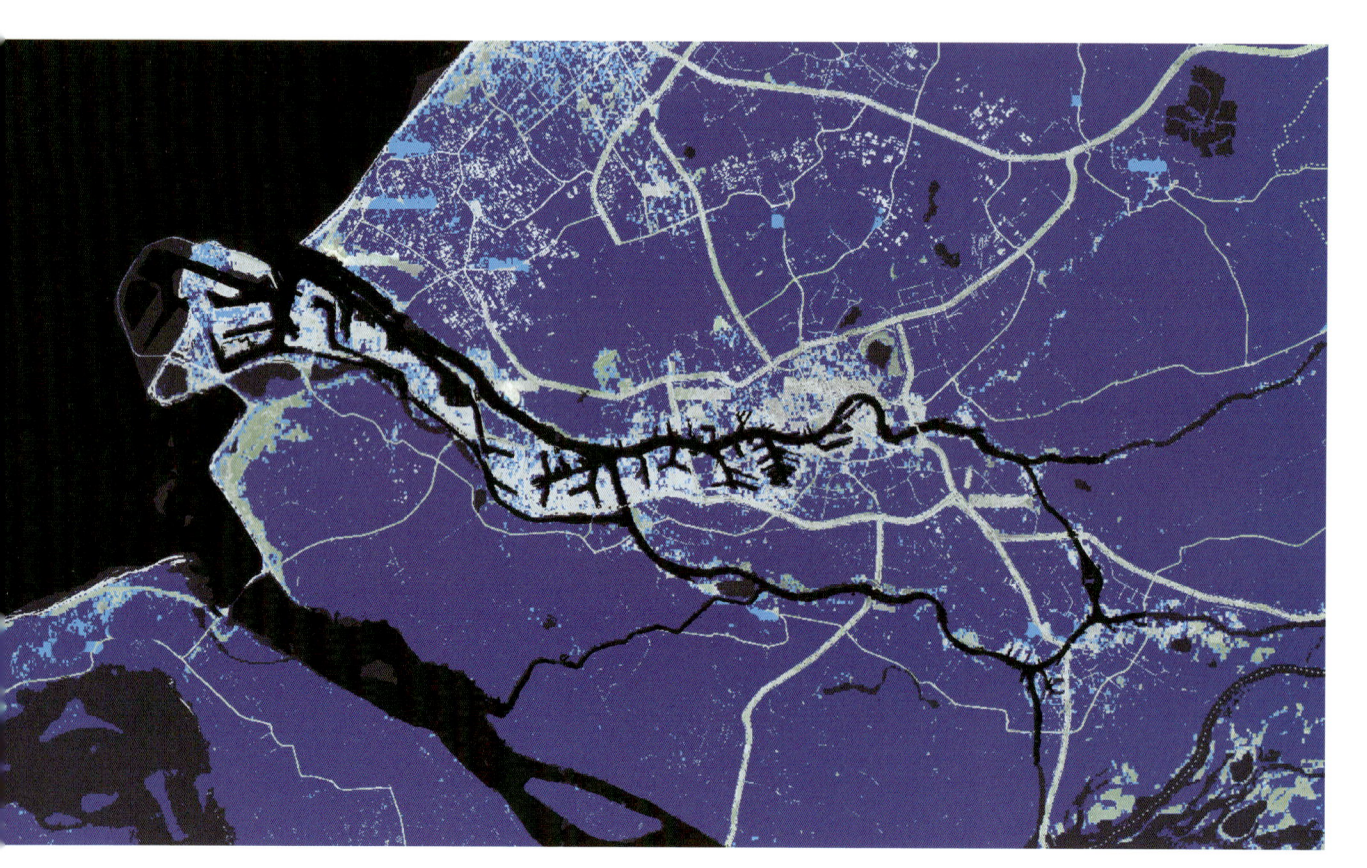

Rotterdam flood map
(Adapted from: Climate Central, Surging Sea-Risk Zone Map, http://sealevel.climatecentral.org/)

Wet Flood Proofing

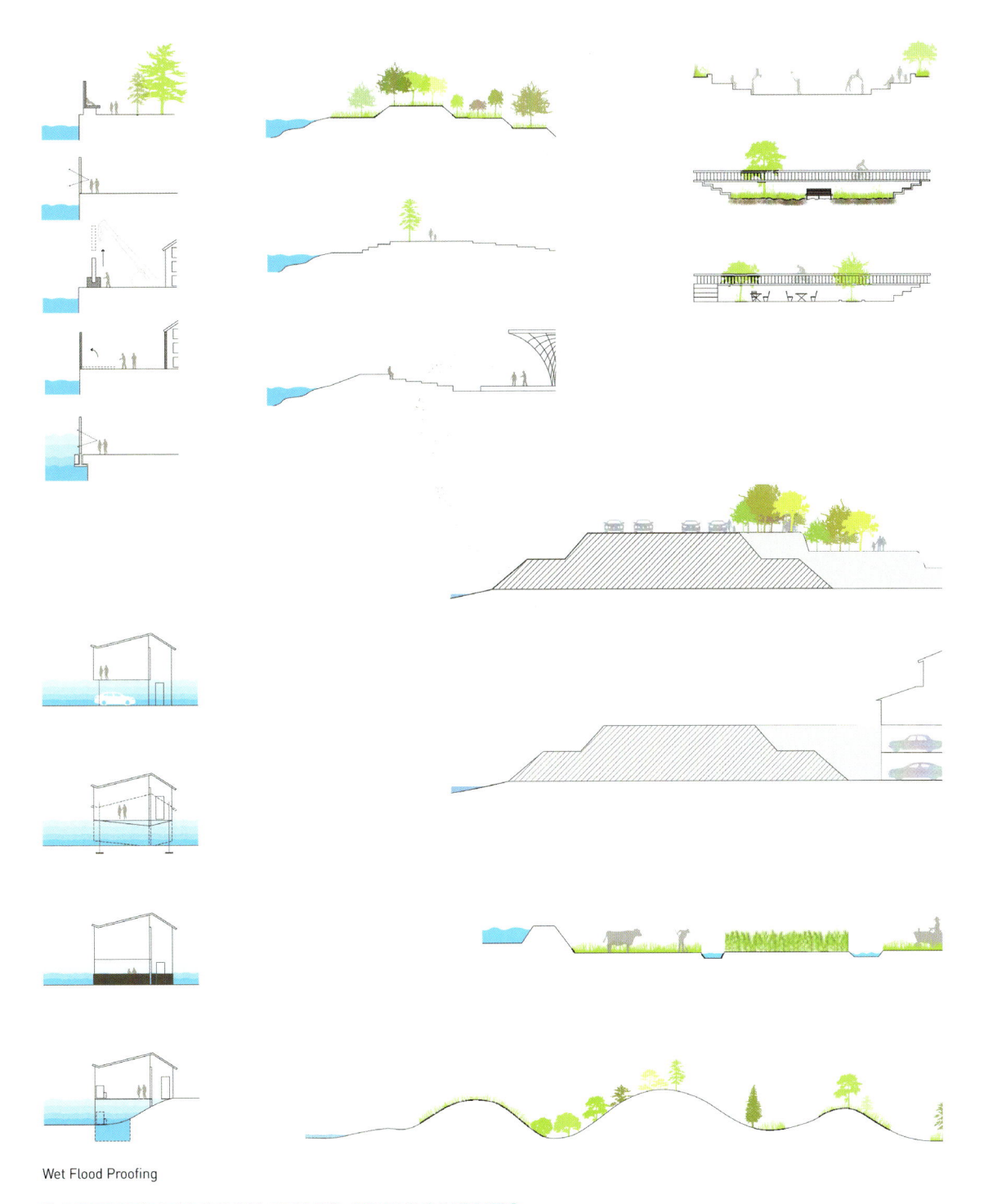

ROTTERDAM RESILIENCE COMPONENTS:
Integrated improvement, dike, water square, multipurpose dike, building codes, polder, dune

Maeslant Barrier, Rotterdam, the Netherlands
The barrier is normally open to allow ships to pass but closes to protect Rotterdam from storm surges. Part of the Delta Works, it is one of the largest moving structures in the world, with the closure trusses alone measuring 780 feet (238 m).

plan exists in the larger context of the dunes, which by far constitute the vast majority of the Netherlands flood protection system. Finally, in addition to these tried and tested gray and green strategies, a new strategy is Building with Nature, such as the Sand Engine, which strengthens shorelines using natural systems. In short, the Dutch protection strategy consists of dunes for the most part, the Delta Works where there are no dunes, Building with Nature pilot projects, and finally, city-level water management strategies, such as Rotterdam's plan.

Rotterdam is becoming a global leader in climate change adaptation and water management. The city aims to be entirely "climate proof" by 2025, the year when the majority of the world's population will be living in flood-prone, low-elevation coastal zones—delta cities just like Rotterdam. All of Rotterdam's experiences are shared in the Connecting Delta Cities project of C40 Cities, a leadership group that exchanges

Rotterdam resilience plan
Rotterdam is a living laboratory of city climate proofing, from water squares and dikes to green-roof programs and floating buildings.
(Source: Rotterdam Center for Resilient Delta Cities)

best practices and climate adaptation strategies among more than ninety delta cities, including New Orleans, New York, and Ho Chi Minh City (whose resilience plans are featured in the following chapters), as well as cities such as Tokyo, Hong Kong, London, and Jakarta. The city has also established Flood Control 2015, a public-private consortium that develops "smart" flood control systems. For instance, it develops "smart dikes" with sensors that provide real-time flood information to a central crisis center, which can relay this information to apps to tell citizens where to go to evacuate safely in times of crisis.

Rotterdam's resilience plan divides the city into districts based on their density, age, and relative position to the existing inner-ring dikes. Newer districts on the periphery of the city will get larger-scale investments in green infrastructure to manage stormwater, elevate particular areas of land, and raise the water table. Older districts with denser building fabric within the older inner-dike system will have large-scale building integrated with

National Water Centre, Rotterdam
This floating pavilion consists of geodesic domes made of steel and ETFE, a transparent, lightweight foil membrane one hundred times lighter than glass.
(Source: Guilhem Vellut, Flickr)

dike reinforcements, and in some places existing dikes will be replaced with seawalls and other features that will allow for more development. Older port facilities in the city center, known as the Stadshavens district, will see a combination of dike reinforcement, dike redevelopment with building integration, and large-scale green infrastructure.

The larger philosophy of the plan is to consider climate change not as a threat, but as an opportunity to make Rotterdam a more attractive city to live, work, and play in, as well as to invest in. The plan addresses climate change in a wide variety of ways, such as water plazas and an extensive green-roof program, that beautify the city but also help store stormwater. There are even floating and amphibious homes and structures, which house entire communities. The National Water Centre is a floating pavilion that is largely self-sufficient, with its own wastewater purification system, solar collectors, heat recovery system, and adjustable microclimate zones. The pavilion is meant as a showcase of new technologies that positions Rotterdam on the cutting edge of flood management and climate change adaptation.

STADSHAVENS DISTRICT

Stadshavens is a port area in transition in the middle of the city of Rotterdam. Major port activities are moving to new districts downriver as Stadshavens develops into a new commercial and residential district, redeveloping almost 1,619 hectares (4,000 ac). The plan calls for reinforcing existing dikes and expanding them into areas in which new property development is integrated into the profile of the dike itself, both to transition more gradually in elevation and to help pay for the dike reinforcement. It also calls for the conversion of port areas outside the existing or new dike perimeter into floodplain parks. The plan includes a research, design, and manufacturing campus, a former shipyard consisting of industrial and dock buildings that now houses education and business development for the building, logistics, and energy industries.

POSTWAR DISTRICTS

The outer districts of Rotterdam, generally postwar urban districts, are dispersed and lower density than the rest of the city, and are generally prone to both flooding and drought. The resilience plan calls for expansion of a park and ecological network that can absorb rain and floodwater, leading to groundwater recharge. These blue-green ribbons, in which landscape features that absorb rainwater are integrated with park and recreation space, will be supplemented by larger scale floodplain parks and agricultural areas.

INNER-DIKE DISTRICTS

The inner-dike districts are already dense, desirable residential districts, so most of the interventions here are proposed at the block or the street level. In most cases, the recommendations are for using apartment blocks' inner courtyards for rainwater capture and infiltration, to lessen pressure on the flood infrastructure, and for extending the canal system for the same purpose.

OUTER-DIKE DISTRICTS

Because of their proximity to the water, the outer-dike districts are some of the most desirable neighborhoods in Rotterdam. But they lie outside the main protective ring of dikes, making them more vulnerable to flooding. Most protective measures call for integrating floodwalls with new waterfront park development, or for removable floodwalls that can be installed in advance of a storm. In some cases, integrating flood protection with stepped waterfront plazas will be enough, but local and building-scale flood protection should also be considered.[2]

CHAPTER 3:
NEW YORK CITY, NEW YORK, USA

New York, vision for 2050
(Adapted from: *PlaNYC, A Stronger, More Resilient New York*, 2013, New York City Special Initiative for Rebuilding and Resiliency, http://www.nyc.gov/html/sirr/html/report/report.shtml)

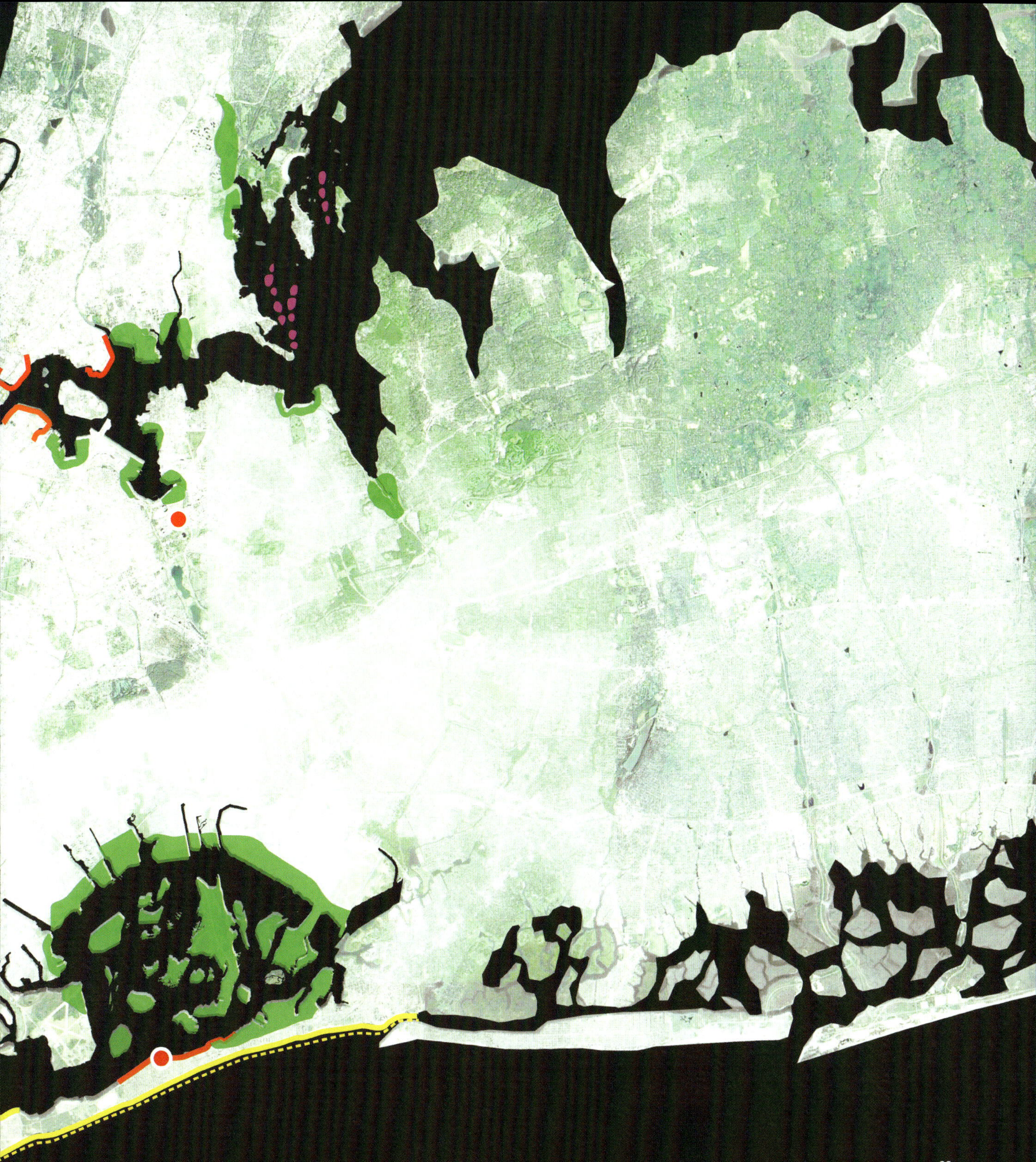

▶ New York City, the most populous city in the United States, has developed multiple resiliency plans over the past decade, beginning with Mayor Michael Bloomberg's PlaNYC in 2007. Up until Superstorm Sandy in 2012, however, flood protection measures were inadequate, especially considering the trillions of dollars of assets at risk.[1]

Following Sandy, which led to the tragic death of forty-three people, New York City updated PlaNYC to incorporate initiatives for urban resiliency—to combat flooding, storm surge, and sea level rise.[2] In 2015, Mayor Bill de Blasio released OneNYC, a plan focusing on large-scale resiliency and building upon the initiatives laid out in PlaNYC's 2013 release. As a result, New York City has developed a fairly robust approach to build resiliency by focusing on critical assets while studying longer-term, more-regional initiatives. Although not as comprehensive and integrated as Rotterdam's, New York City's plans certainly set a new standard for U.S. cities.

An early plan called for regional barriers for flood protection, including the New York–New Jersey Outer Harbor storm surge barrier, similar in strategy to New Orleans. But for political and other reasons, the barrier idea was abandoned in favor of a plethora of local strategies, including multipurpose levees and building codes, which will end

New York flood map
(Adapted from: Climate Central, Surging Sea-Risk Zone Map, http://sealevel.climatecentral.org/)

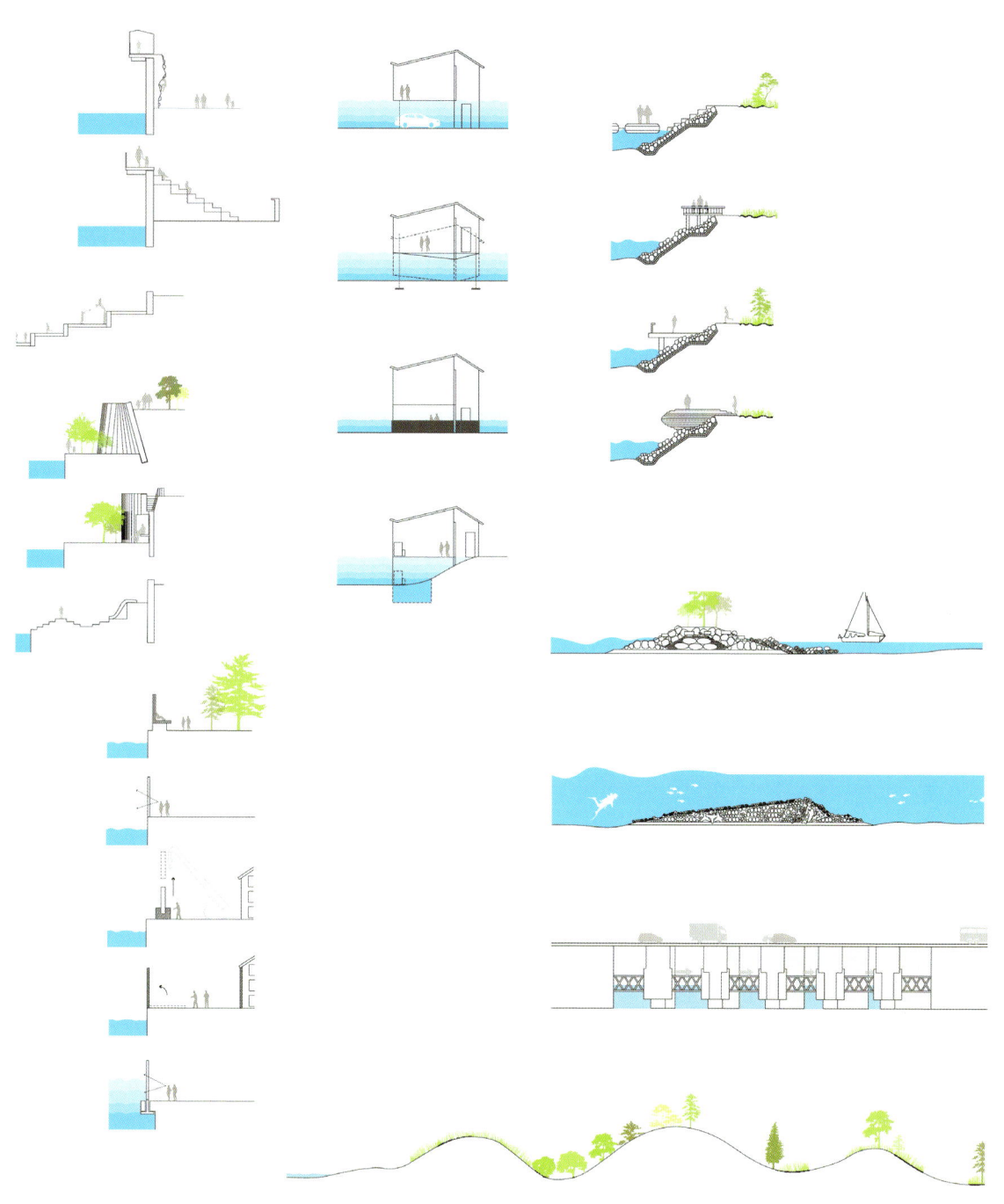

NEW YORK RESILIENCE COMPONENTS:
Levee/integrated development, building codes, revetment, breakwater, surge barrier, dune

up being more expensive.[3] Nevertheless, the Army Corps of Engineers is still studying the use of barriers, such as an East Rockaway barrier, as a first line of defense. Although no plans for implementation currently exist, it would bring local resilience initiatives to a halt.

New York City joined the Rockefeller Foundation's 100 Resilient Cities program in 2013. This program has a view of resilience that does not just include the shocks—superstorms, blackouts, heat waves, and other acute events—but also the stresses that weaken the fabric of a city on a day-to-day basis, such as high unemployment.[4] The program cosponsored, in partnership with the U.S. Department of Housing and Urban Development, the Rebuild by Design competition. It led to several more detailed design interventions, including the iconic BIG U for Lower Manhattan, spearheaded by the designers of the Bjarke Ingels Group.

PlaNYC is structured around four coastal protection strategies: 1) increase the elevations of the coastal edges, 2) minimize the upland wave zones, 3) protect against storm surge, and 4) improve coastal governance and design.

INCREASE COASTAL EDGE ELEVATIONS

To reduce the risk of tidal flooding as a result of sea level rise by the 2050s, the city is planning to increase the height of vulnerable coastal edges. Measures include beach nourishment and bulkheads.

MINIMIZE UPLAND WAVE ZONES

The city has a range of projects aimed at diminishing the force of storm waves, both up shore and onshore, before they can damage urban areas. Initiatives include dune construction and beach nourishment in the Rockaway Peninsula, Queens, an area at risk of extreme flooding and wave action. Superstorm Sandy destroyed many homes on Breezy Point on the Rockaway Peninsula. The Army Corps of Engineers has already built 6 miles of dunes and replenished beaches there with 3.5 million cubic yards (2.7 million m^3) of sand. A project that includes levees and flood walls is currently seeking financing.

The strategy of minimizing upland wave zones also includes several offshore breakwater projects, including around Great Kills Harbor, Staten Island, an area vulnerable to wave action and erosion that can undermine shoreline bluffs and damage homes in surrounding neighborhoods. In addition, this approach features natural strategies such as wetlands and living shorelines. In Tottenville, Staten Island, a living shoreline project is being built that includes oyster reef breakwaters, beach nourishment, and maritime forest enhancements. Living shorelines and floating breakwaters are also planned in Brant Point, Queens. Wetlands are planned to protect neighborhoods around Howard and Hamilton Beaches in Queens.

PROTECT AGAINST STORM SURGE

New York City plans to use flood protection structures, including levees, floodwalls, and local storm surge barriers, to keep water away from critical infrastructure and vulnerable neighborhoods. The plan includes an integrated flood protection system in Hunts Point, the Bronx, the location of a food distribution center that is vital for the city's food supply. This strategy also includes integrated flood protection for East Harlem, including along the FDR Highway, as well as in Lower Manhattan, including the Lower East Side, which has a very large residential population and density, as well as residents with low and moderate income levels. The Lower Manhattan flood protection system would later be dubbed the BIG U, for its U shape embracing the tip of Manhattan. A critical component is the floodwalls at Hospital Row, which is prone to major flooding, to keep hospitals open during future storms.

The BIG U

The BIG U, designed by the Bjarke Ingels Group and One Architecture, is conceived as a 10-mile (16-km) loop around Manhattan with various design approaches to resilience, ranging from deployable walls to waterfront parks, such as a bridging berm at East River Park that protects from floods, offers recreation, and provides accessible routes over the highway.
(Source: The Bjarke Ingels Group)

IMPROVE COASTAL DESIGN AND GOVERNANCE

The final part of the New York resilience strategy is focused on using natural areas and open space to protect from floods, as well as on improving the management of waterfront assets and permitting. It includes citywide waterfront inspections of the city's coastal assets, providing design guidelines for waterfront areas, and streamlining the in-water permitting process, as well as promoting green infrastructure and innovative coastal protection techniques.

CHAPTER 4:
NEW ORLEANS, LOUISIANA, USA

New Orleans, vision for 2050
(Adapted from: Greater New Orleans Urban Water Plan, 2013, http://livingwithwater.com/blog/urban_water_plan/plan/; Resilient New Orleans: Strategic Actions to Shape Our Future City, 2015, http://resilientnola.org)

39

▶ The city of New Orleans, located on the southern coast of the state of Louisiana, is a prime example of how sea level rise and flooding can have extreme effects on the operation of a city. New Orleans sits at the mouth of the Mississippi River, which drains about 40 percent of the water in the continental United States. New Orleans is also located on the Gulf of Mexico, which has been the source of flooding for large storms such as Katrina. Additionally, New Orleans is a low-lying city subject to land subsidence due to geologic and human-induced processes. More than one-third of the region's land area is wetlands, and a majority of the land area is at or below sea level.

In 2005, Hurricane Katrina flooded about eighty percent of the city, leading to more than fourteen hundred deaths. The Army Corps of Engineers started an effort to rebuild and raise the city's existing concrete levees and build barriers, including the Gulf Intracoastal Waterway West Closure Complex, a navigable floodgate with the world's largest pump station, and the 1.8-mile-long (2.9-km) Lake Borgne Surge Barrier, designed to protect the east side of the city from storm surge coming from the Gulf of Mexico and the lake. In total, more than $20 billion of local, state, and federal money was spent on 350 miles (562 km) of levees, flood walls, gates, and pumps. The improved system is built to a 100-year flood standard, with a 1 percent chance every year of a flood (compare

New Orleans flood map
(Adapted from: Climate Central, Surging Sea-Risk Zone Map, http://sealevel.climatecentral.org/)

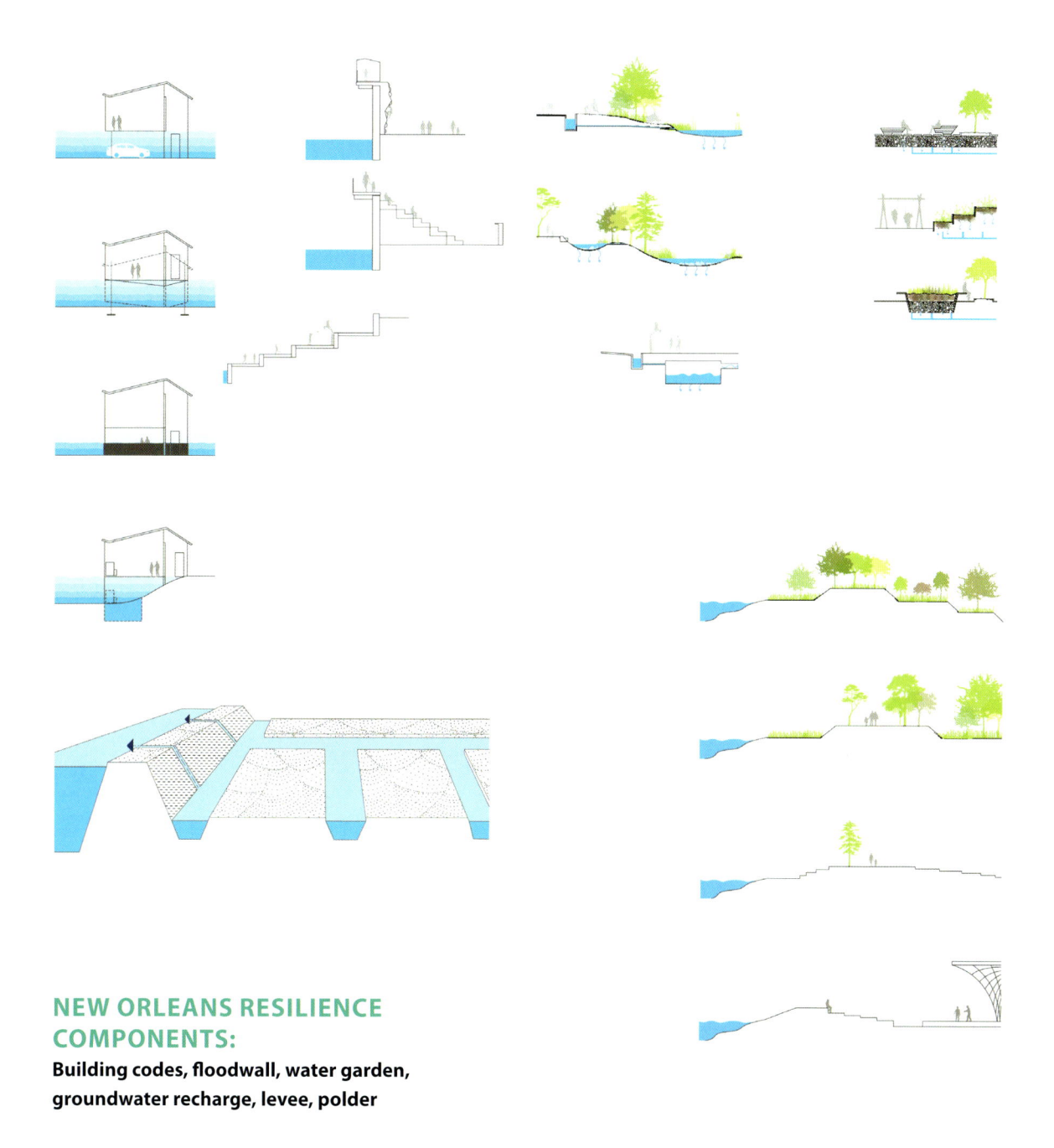

NEW ORLEANS RESILIENCE COMPONENTS:

Building codes, floodwall, water garden, groundwater recharge, levee, polder

to Rotterdam's much safer 10,000-year flood standard). But experts wonder whether this is enough, especially considering climate change and the more powerful storms the city will face, increasing the likelihood of a flood.[1]

RESILIENT NEW ORLEANS PLAN

The Resilient New Orleans plan, guided by the Rockefeller Foundation's 100 Resilient Cities program, brings together local understanding and global best practices to reduce risk and inequity.

Resilient New Orleans focuses on the effects of large storms as well as the daily stresses on the natural and built environment. In particular, Resilient New Orleans looks to end wetlands loss and stabilize and expand the city's flood protection systems. New Orleans is particularly concerned with sea level rise and coastal flooding as it relates to the issues of poverty and resource allocation within the city, as much of the at-risk population for flooding is poorer, with higher unemployment and distress in their communities. Resilient New Orleans pays particular attention to the needs of these populations.

The report has three distinct sections: Adapt to Thrive, Connect to Opportunity, and Transform City Systems.[2]

ADAPT TO THRIVE

This vision highlights New Orleans's multiple lines of defense approach. This includes hard infrastructure solutions paired with soft infrastructure, regulations, planning, and investment. The plan will advance coastal protection and restoration. This includes supporting the efforts of the Louisiana Coastal Protection and Restoration Authority and leveraging the critical resources to accomplish coastal protection projects, including the restoration of coastal wetlands, which provide flood protection, and workforce opportunities.

In addition, the plan calls for investing in innovative and comprehensive urban water management. This includes the Greater New Orleans Urban Water Plan, an ongoing project to promote the vision of a protected city, which includes adding green infrastructure to complement the traditional drainage system. The plan also helps property owners invest in risk reduction through stormwater infrastructure by offering incentives as part of a "resilience retrofit" program offering loan repayment through property tax bills and a potential reduction in insurance premiums.

An important part of the vision is the creation of a culture of environmental awareness. The establishment of a resilience center, led by the City of New Orleans and supported by the Rockefeller Foundation and 100 Resilient Cities, will promote outreach and capacity building of the Office of Resilience and Sustainability. Additionally, the city launched a coastal design public awareness campaign, increasing community discussion and understanding around the threat of flooding and sea level rise, and

Lake Borgne Surge Barrier, New Orleans

The Inner Harbor Navigation Canal Lake Borgne Surge Barrier is 28 feet (8.5 m) high, with pilings reaching up to 100 feet (30.5 m) into the lake.

(Source: US Army Corps of Engineers, Wikimedia Commons)

Lafitte Blueway, New Orleans
The proposed Lafitte Blueway, here represented during dry conditions, includes areas for storing water along the river.
(Source: Bosch Slabbers, Greater New Orleans, Inc., 2010, "Greater New Orleans Urban Water Plan")

individual mitigation options. Finally, the plan supports a commitment to mitigating New Orleans's climate impact, with greenhouse gas reduction targets for 2050.

CONNECT TO OPPORTUNITY

This vision acknowledges the importance of equity as an engine for economic, social, and environmental resilience, hence there are several plans dedicated to improving equity. The plan includes investing in household financial stability, for instance, by setting up emergency savings accounts. Other components of the plan are lowering the barriers to workforce participation, such as, through the promotion of digital literacy; promoting equitable public health outcomes, such as by increasing access to fresh food in traditionally underserved neighborhoods; continuing to build social cohesion, such as via a program aimed at reducing gun violence; and expanding access to safe and affordable housing.

TRANSFORM CITY SYSTEMS

The final aspect of the plan is focused on improving the city's infrastructure to promote economic development and improve resilience. It includes the redesign of an integrated and efficient regional transit system to connect people on a regional level. Promoting energy efficiency, redundancy, and reliability is also a key aspect, through an energy efficiency challenge for businesses, for instance, or the installation of backup generators. Finally, the plan aims to integrate resilience-driven decision making across various public institutions and to develop the disaster preparedness of businesses and neighborhoods through capacity-building programs.

CHAPTER 5:
HO CHI MINH CITY, VIETNAM

Ho Chi Minh City, vision for 2100
(Adapted from: Climate Adaptation Strategy Ho Chi Minh City, 2013, Vietnam Climate Adaptation Partnership (VCAPS) consortium, City of Rotterdam, and Ministry of Infrastructure and Environment, The Netherlands, http://www.vcaps.org/assets/uploads/files/HCMC_ClimateAdaptationStrategy_webversie.pdf)

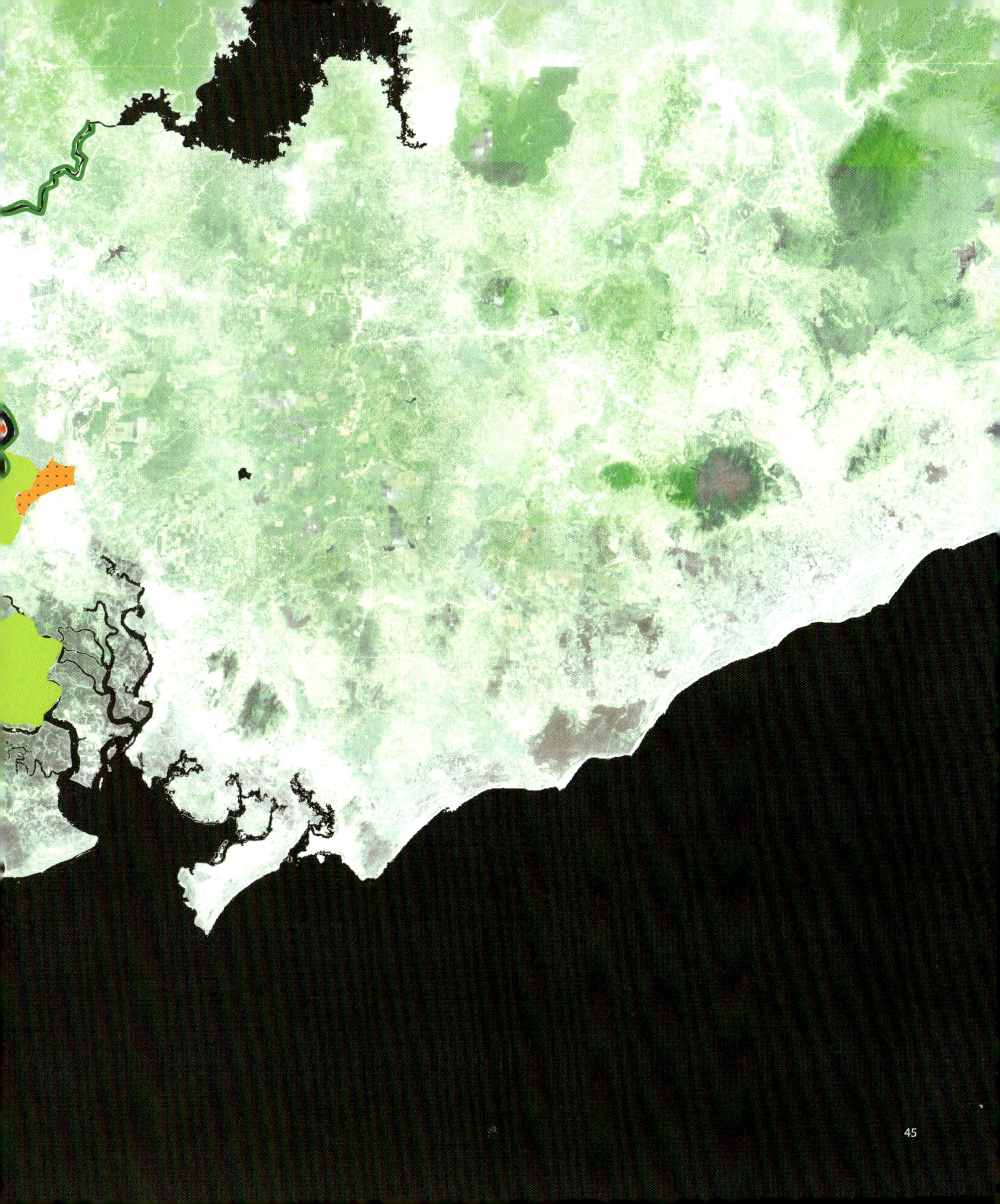

▶ Explosive population growth in Ho Chi Minh City, Vietnam's largest city, has brought the population to more than 8 million people. This rapid growth has led to building in floodplains and the region's most ecologically sensitive areas, as well as a strain on the city's infrastructure. The climate change threats to Ho Chi Minh City include a rise in temperature and a heat island effect, potentially causing reduced air and water quality, increased flooding events and saltwater intrusion as a result of sea level rise, and changes in precipitation and river runoff. The impacts of these changes are a threat not only to Ho Chi Minh City residents, and in particular to the city's vulnerable "floating population" of an estimated 2 million migrant workers, but also to the millions of people in the hinterland who depend on the city for resources. There are risks to the economy of Southeast Asia if Ho Chi Minh City and its large port are unable to function at its current capacity.

Because Ho Chi Minh City and Rotterdam are both deltas of large riverine systems and large port cities dependent on waterfront access to sustain their economies, the two cities are in cooperation with one another through the Connecting Delta Cities program, a project of C40 Cities. Borrowing from Rotterdam, Ho Chi Minh City is now taking an integrated approach to flood protection, using flood management as a way to increase the attractiveness of the city, improving both

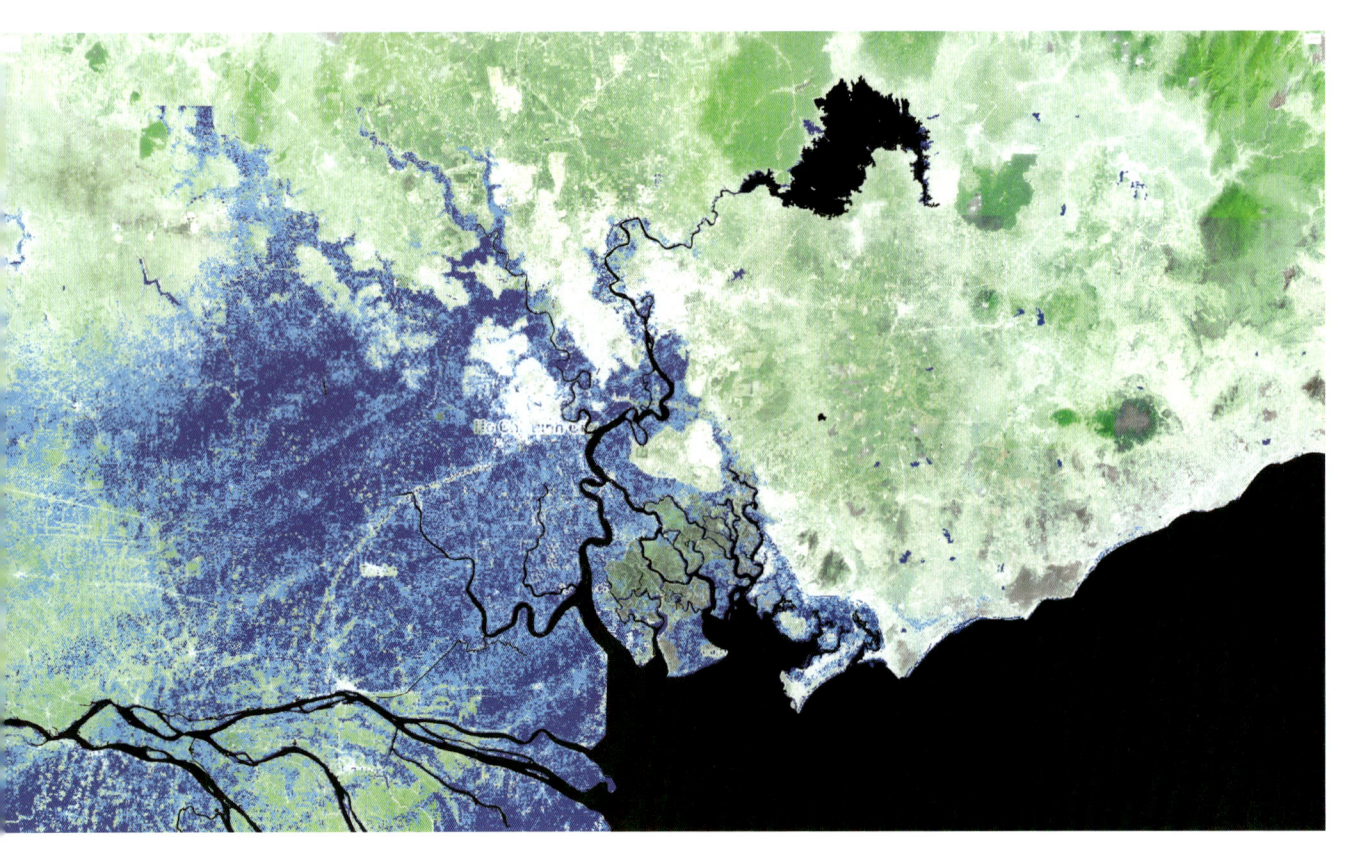

Ho Chi Minh City flood map
(Adapted from: Climate Central, Surging Sea-Risk Zone Map, http://sealevel.climatecentral.org/)

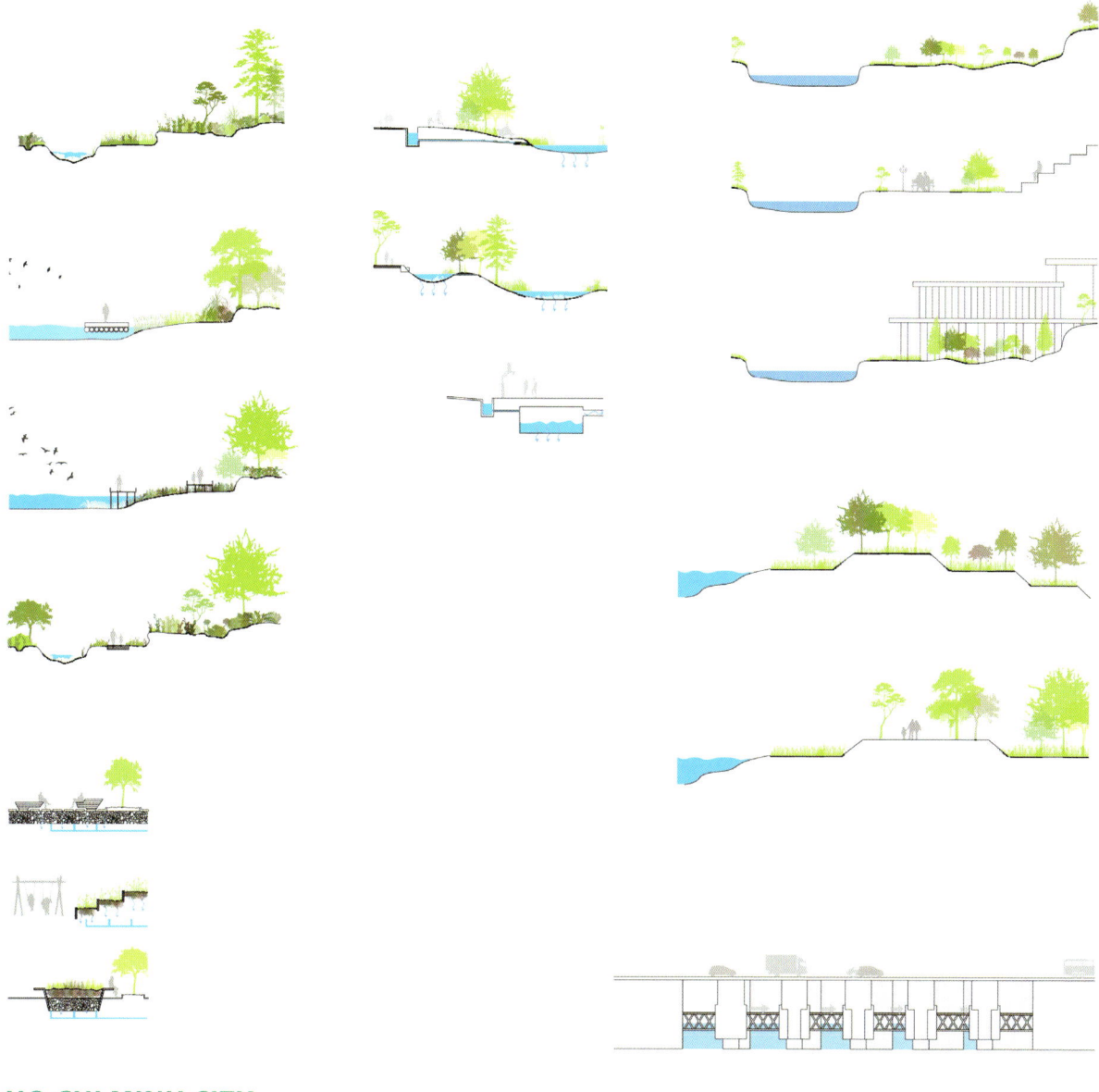

HO CHI MINH CITY RESILIENCE COMPONENTS:
Biofiltration, water garden, floodplain park, groundwater recharge, dike, surge barrier, multipurpose dike

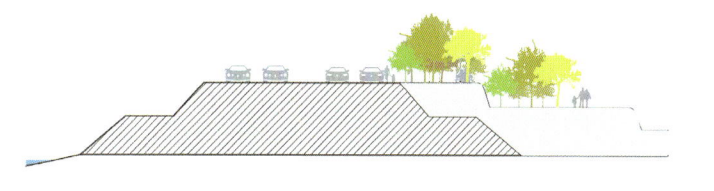

Climate-proof plan for District 4, Ho Chi Minh City

The plan for District 4, where Saigon Port is located, includes a waterfront edge with iconic bridges and buildings, a large stepped levy with an underground road, the development of a market street with underground water storage, and green arteries along roadways.

(Source: Rotterdam Center for Resilient Delta Cities, http://rdcrotterdam.com/projects/ho-chi-minh-city-moving-towards-the-sea-with-climate-change-adaptation/)

safety and socioeconomic development goals. The strategic directions include basing urban development on soil and water conditions, using a stepwise and multiscalar approach that includes both regional and local flood measures, increasing the city's water storage and drainage capacity, building a citywide network of green infrastructure, and implementing green building codes to reduce heat island effects.[1]

First and foremost, to respond to the effects of sea level rise and climate change, Ho Chi Minh City is developing primary infrastructure such as dikes. However, its low-lying and delta location, as well as rapid growth and development, present challenges. The city is integrating dike developments in floodplains with new urban districts.

In the city's outer districts, which are still in the process of urbanizing, strategic retreat is a viable option.

PROTECTIVE RING DIKE AND DIKE REDEVELOPMENT

Ring dikes are being developed to protect the built-up parts of the city, which have the most capital at risk from a potential flood. These are integrated with a ring road. Outdated harbors near the city center are to be redeveloped with integrated flood protection, including multipurpose dikes and stepped dikes that integrate floodplain parks with urban development. Because the Saigon River runs through the middle of the city, the plan aims to keep the river floodplains free from development to increase the river's capacity.

ADAPTING BUILDINGS OUTSIDE OF THE RING DIKE

In areas outside of the ring dike, smaller-scale adaptive measures are to be implemented. These include raising buildings above the level of the floodplain and wet-proofing new buildings—allowing water to move in and out of buildings during times of flooding while minimizing damage, for instance, by anchoring the structure, using flood-resistant materials, and protecting a building's mechanical and utility infrastructure. New development is to be concentrated on high ground or on mounds of fill (for developments smaller than 150 hectares (370 ac)) or protected by local ring dikes (for developments larger than 150 hectares (370 ac)). Development in areas that are most vulnerable should strategically retreat to these new development areas as necessary.

LANDSCAPE INFRASTRUCTURE

Increasing salinity due to sea level rise means more salt-tolerant vegetation is needed, because less salt-tolerant vegetation may not survive. With a high level of impervious surfaces, Ho Chi Minh City looks at ways of promoting stormwater capture and infiltration, especially within the ring dike, to get rid of excess water quickly. Increasing tree and park cover can lessen heat island effects and provide recreational space for residents.

PART II
LOCAL STRATEGIES

Chapter 6: Hard-Protect Strategies

Chapter 7: Soft-Protect Strategies

Chapter 8: Store Strategies

Chapter 9: Retreat Strategies

Chapter 10: Conclusion

CHAPTER 6: HARD-PROTECT STRATEGIES

▶ The first section on local solutions features the hard-protect strategy. "Hard" here means an engineered, gray solution, although opportunities for hybrid gray-green projects exist. "Protect" refers to a defense strategy that aims to keep the water out, either by holding the line of defense ("defend"), or by aggressively advancing the line through land reclamation ("attack"). The hard-protect strategy has been, and to a large extent still is, the dominant method of flood protection, although soft-protect solutions, as well as store and retreat strategies, are increasingly taking hold. Strategies vary from large regional engineered solutions to smaller-scale and building-level solutions.

Examples of hard-engineered protect strategies featured in this section include seawalls, revetments, breakwaters, floodwalls, dikes (or levees), and surge barriers. Traditionally, the sole purpose of these infrastructure investments is to protect from flood risk, but opportunities exist to make them multifunctional by integrating them with walkways, public plazas, parks, and buildings. In this way, they could have wider social, environmental, and economic benefits beyond protecting from floods alone, for instance, by unlocking the real estate and economic development potential of the newly secured areas, which could help pay for the infrastructure. In short, hard-protect infrastructure should not only protect from floods, but should be integrated into the landscape, buildings, and the public realm.

6.1. Protect + Reoccupy / Reclaim

6.2. Seawall

6.3. Revetment

6.4. Breakwater

6.5. Floodwall

6.6. Dike

6.7. Multipurpose Dike

6.8. Surge Barrier

6.1 PROTECT + REOCCUPY/RECLAIM

Protect and reoccupy/reclaim is a strategy that capitalizes on the increased real estate value of a new protect measure, such as a dike, which lowers the flood-risk levels of land and hence increases property values. It does so by building either behind the existing line of defense after a relocation (protect and reoccupy), or in front of the existing line of defense after reclamation (protect and reclaim). Because waterfront properties often have high development potential, the large investment in shoreline fortification can be feasible when paired with redevelopment projects. The construction cost for large infrastructure, dredging, diking, fill, and land elevation can be paid by the additional real estate value that is created.

Protect and reoccupy is a "hold the line" strategy of defense. Protect and reclaim is an "attack" strategy that goes seaward. It has been used in places with high population densities and land scarcity, such as Hong Kong, Singapore, and the Netherlands, although initially not deliberately as a measure against sea level rise.

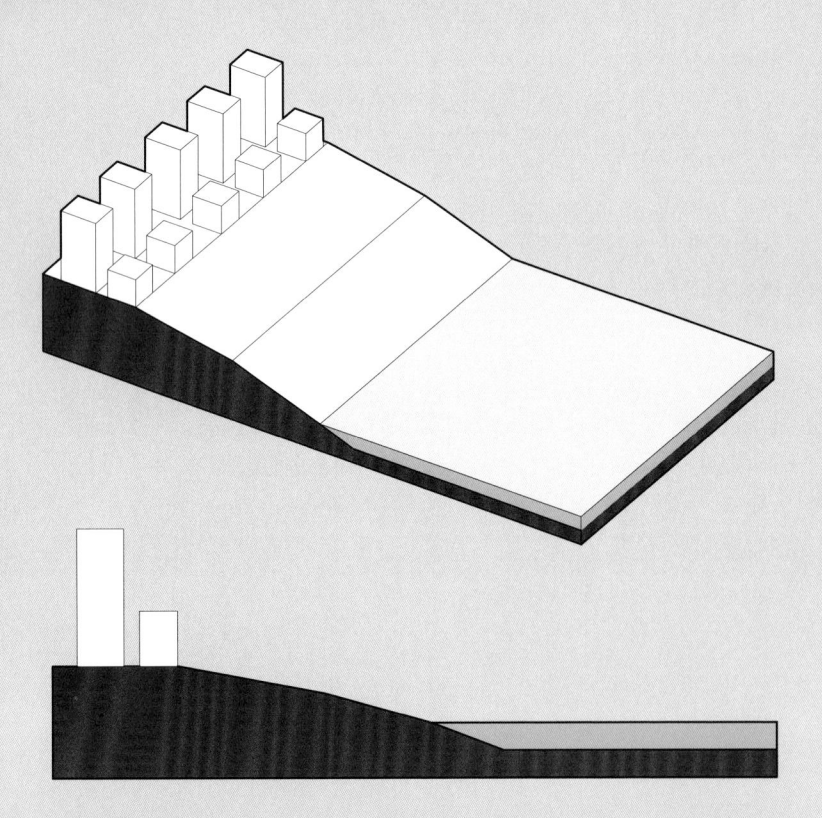

PERFORMANCE Protect and reoccupy/reclaim provides an opportunity to deal with sea level rise on a longer-term basis. In addition, it capitalizes on the value of the infrastructure created (the reduction of flood risk) by adding real estate.

PROS

⊕ Can be a longer-term solution for sea level rise

⊕ Cost for infrastructure can be offset by the value and the tax revenue created from the new development

⊕ Can create space where there is a constraint (for protect and reclaim)

DESIGN GOALS

◉ Minimize environmental cost

◉ Establish new network of connections and transportation

◉ Provide access and views from new districts and dikes to water

CONS

⊖ Expensive to build infrastructure and to elevate land

⊖ Large operational and maintenance costs

⊖ Relocation of residents (if necessary) can be a challenge

⊖ Increased development density also increases risk of dike/dam failure

⊖ Possible environmental cost of reclamation, particularly of draining wetlands or infilling reefs, which destroys habitats (for protect and reclaim)

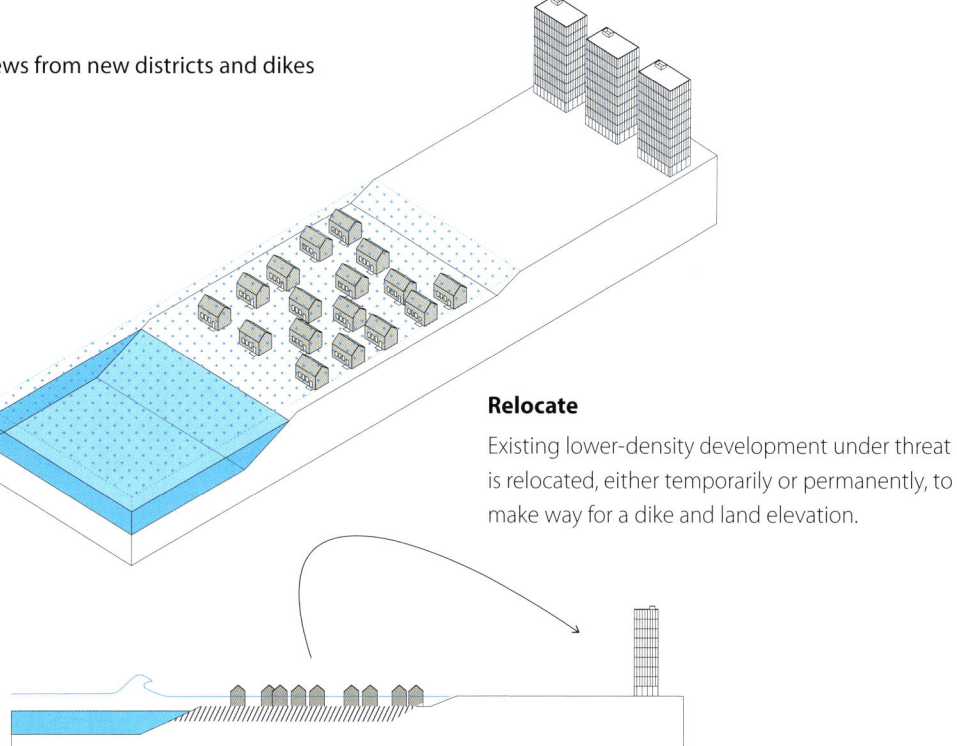

Relocate

Existing lower-density development under threat is relocated, either temporarily or permanently, to make way for a dike and land elevation.

PART II: HARD-PROTECT STRATEGIES | 55

Protect

Once existing settlements are temporarily or permanently relocated, a new dike system and land elevation protect the area behind the dike.

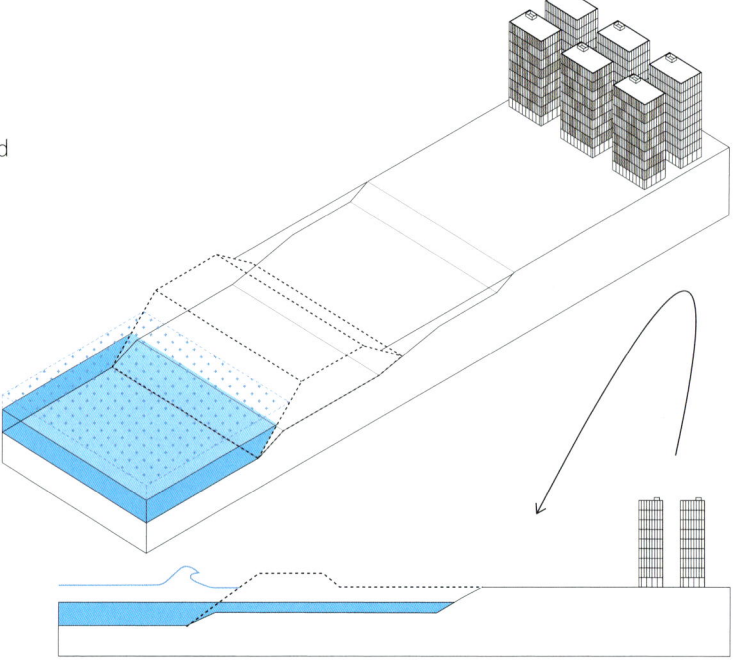

Reoccupy

The elevated land with a fortified edge is now suitable for new development. The costs for infrastructure and land elevation are compensated by the real estate development revenue.

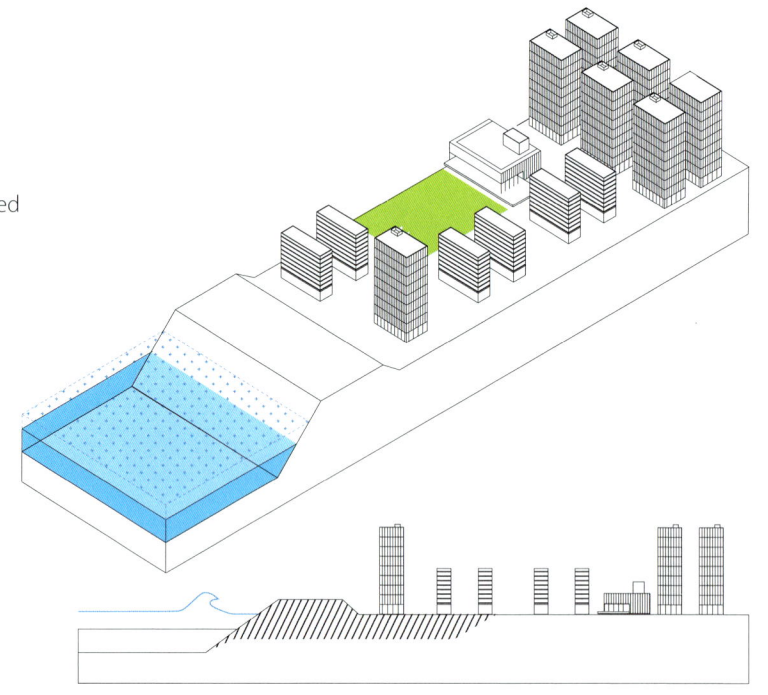

NATIONAL CAPITAL INTEGRATED COASTAL DEVELOPMENT PROGRAM, JAKARTA, INDONESIA

Jakarta's biggest threat is not sea level rise but groundwater extraction, which causes the city to sink and raises the risk of floods. In some areas, the city sinks up to 25 cm (close to a foot) per year. A proposed 40-kilometer-long (25-mi) sea dike across Jakarta Bay (initially known as the Great Garuda for its shape, which was inspired by a mythical bird that is Indonesia's emblem) would protect the city from floods but also create a new lagoon with future energy and aquaculture potential and public transport connections that would contribute toward alleviating Jakarta's infamous traffic congestion. Importantly, it would also offer enhanced locations for new fishing harbors linked to urban revitalization programs and new locations for additional mangrove and coral habitats, while future development opportunities can arise with land reclamation connected to the realization of the flood defense engineering measures. The project is developed by a Dutch-led consortium in collaboration with Indonesian and Korean partners.

National Capital Integrated Coastal Development program

Since Jakarta is constrained for space to build flood protection onshore, it plans to build a seawall across Jakarta Bay while reclaiming areas for new urban development.

(Source: KuiperCompagnons/NCICD2 consortium)

6.2 SEAWALL

Seawalls are vertical structures designed to protect habitation from major wave and tidal action. These structures, which are generally made of concrete but also of stone, create a stark boundary at the shoreline and help prevent upland erosion and storm-surge flooding.

Although seawalls are effective barriers, they can obliterate the relationship between water and land and cause erosion by disrupting sediment movement and often require beach nourishment as beaches disappear. To better integrate seawalls, they can be designed to be multipurpose, with general seating, recreational activities, and beach access in mind, for instance, as climbing walls or amphitheaters. Various public spaces can be created behind the seawall, such as boardwalks, plazas, parks, and other spaces that support commercial activity. These could make the edge of the seawall an interesting and dynamic experience, rather than something with all the charm of a bare concrete wall.

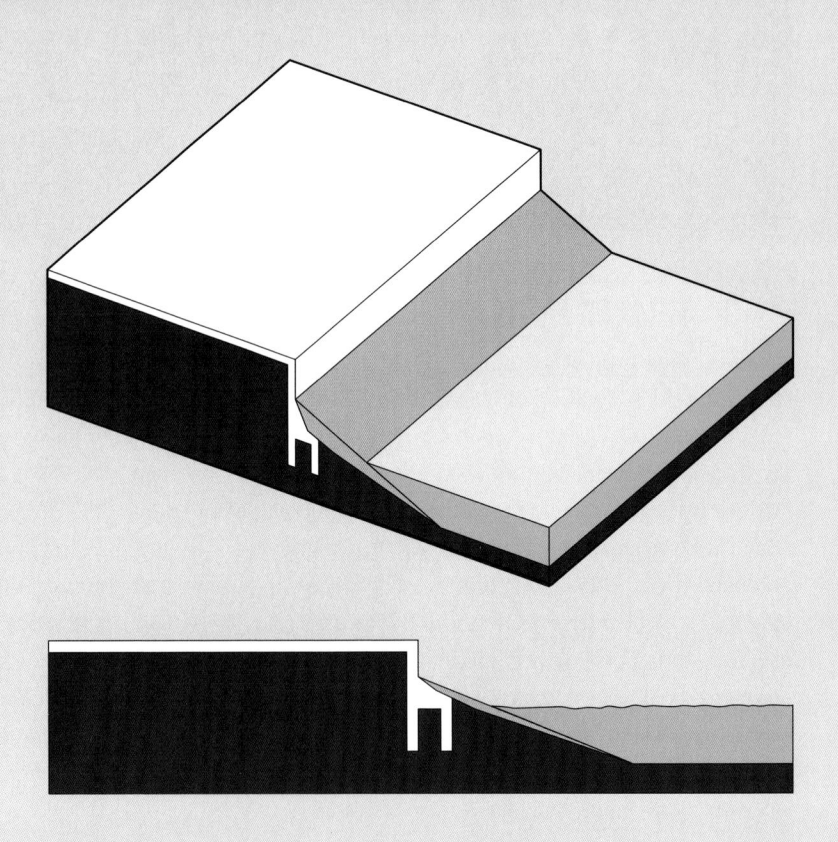

PERFORMANCE Seawalls, a type of floodwall for coastal areas, are vertical barriers built to prevent flooding in upland areas. In contrast to other type of floodwalls, sea walls are found along the coastlines of large bodies of water, such as oceans or seas. Although they have a smaller footprint than other flood protection measures, they alone are not sufficient to meet all needs of flood protection. Unfortunately, they can cause sediment movement as well as ecological and environmental harm. When waves strike a seawall, wave energy is reflected back and carries sand offshore. If the sand is not replenished, the wall will eventually be undermined and fail. Another problem of seawalls is that they can cause erosion to unprotected areas immediately adjacent to the seawall, because the walls reduce the natural replenishment of these areas by impacting littoral drift—the transport of noncohesive sediment, such as sand, along the foreshore.[1]

PROS
- Long-term solution compared to soft solutions such as beach nourishment
- Forms a hard and strong coastal defense that effectively minimizes property damage and loss of life during extreme weather events
- Can be created as multipurpose infrastructure to include public space and commercial activity

CONS
- Very costly to build
- Can cause beaches to vanish
- Can be an "eyesore" in the landscape
- Can alter natural shoreline processes and devastate habitats such as intertidal beaches
- Alters sediment transport processes and disrupts sand movement, potentially leading to increased erosion further down drift

DESIGN GOALS
- Activate edges of engineered infrastructure
- Maintain views toward water
- Create connections between the shore level and the higher top of the seawall

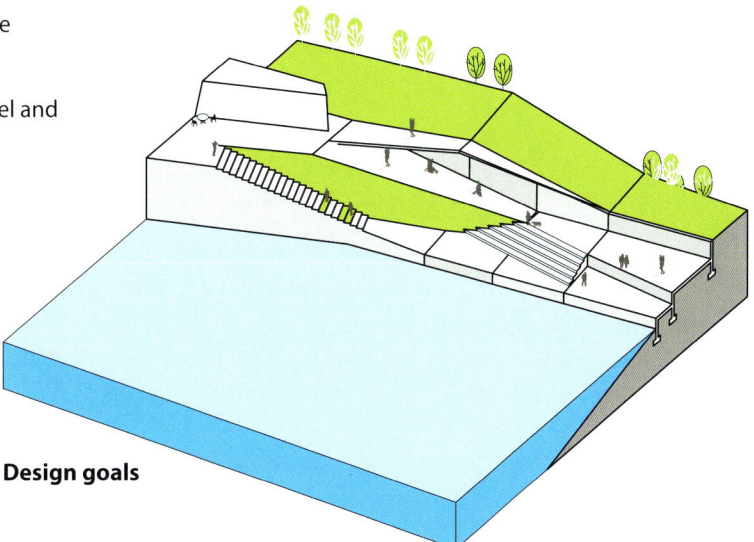

Design goals

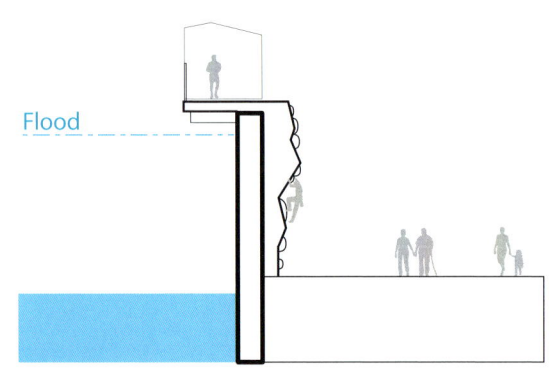

Climbing wall
Activate engineered edge with recreational function as climbing walls.

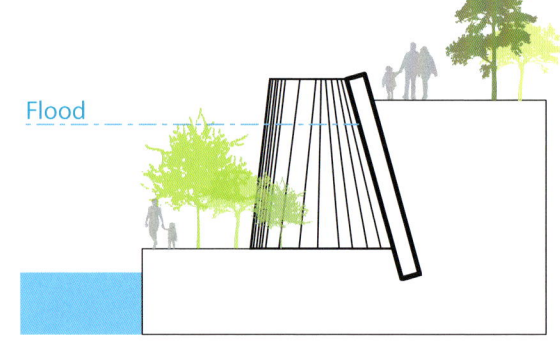

Interweaving ribbons
Overcome elevation change with artfully designed seawall.

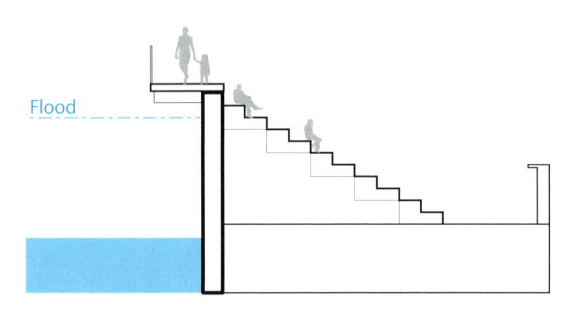

Amphitheater
Overcome height difference with stepped amphitheater.

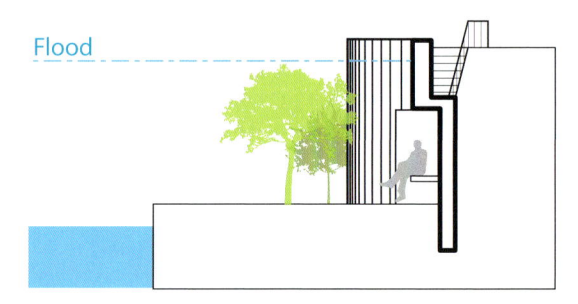

Bench
Incorporate seating area into floodwall.

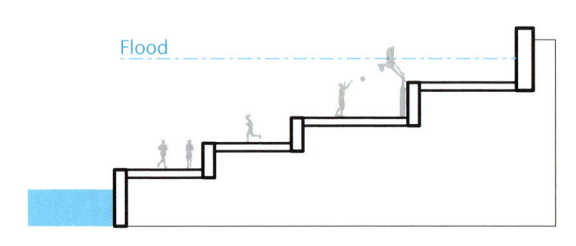

Running trail & park
Program the space between seawall and water.

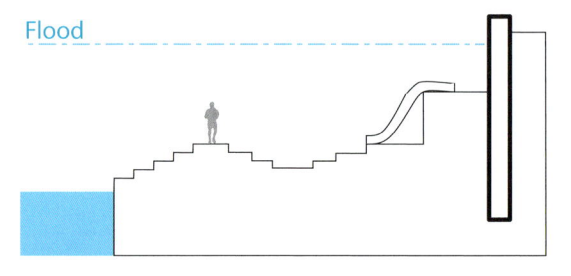

Water plaza
Activate waterfront space and overcome elevation.

THE SEAWALL, VANCOUVER, CANADA

Construction on Vancouver's seawall along its northern shore began in 1917 and was completed in 1980. Most of the seawall is made of stone, such as around Stanley Park, to prevent erosion along the park's shoreline. A 2012 surge storm, combined with high tide, led to significant damage in this area, including rock debris. The seawall's main function is to prevent shoreline erosion from rising sea levels and coastal flooding, but it doubles as a recreational site. Portions of the seawall are multipurpose, including a 28-kilometer-long (17-mi) greenway. There is strong support for the seawall from the city and residents, who enjoy what Vancouver Parks and Recreation claims to be the world's largest uninterrupted waterfront path.

The seawall in Vancouver
Pedestrians stroll along Vancouver's seawall, the city's most popular recreational asset. (Source: Andrew Raun, Wikimedia Commons.)

6.3 REVETMENT

Revetments are flood- and erosion-protection structures on sloped surfaces placed on banks or cliffs. They absorb wave energy, but do not provide protection from storm surge. They are made of wooden planks, concrete blocks, and/or steel, and allow sea water and sediment to pass through. They tend to be lower in cost than vertical walls but take up more space. A beach can build up behind the revetment and provide further protection for the cliff or banks upshore. Riprap revetments, or flexible revetments, are made of mixtures of stone, concrete, and other materials to provide protection from flooding and shoreline erosion.[2]

Revetments on natural sites may lead to loss of habitat. On sandy shorelines, they can accelerate erosion of adjacent sites that are not reinforced. Revetments are appropriate in areas where there is not enough space to do a soft-protect solution, such as a living shoreline. They are increasingly used to stabilize and protect shorelines globally, because they can accommodate some vegetation and can also enable waterfront access.[3]

PERFORMANCE Because the concrete blocks allow for more readjustment than a vertical wall after powerful waves, rip raps are unlikely to fail even when waves damage the structure.[4] Rock riprap is most frequently used, but other riprap types include concrete slab revetments, rubble, and preformed blocks. Usually, the "toe," the end of the revetment toward the sea, is made of concrete or heavy stone to prevent revetment material from sliding.

PROS
- Effective in dissipating wave energy
- May reduce coastal erosion
- Requires less maintenance than a sea wall
- Very long structure life
- Can be extended or modified in the future

CONS
- Typically more expensive than a seawall
- Major landscape impact
- Takes up far more space than a sea wall
- Can alter the dune system as it affects the buildup of sand
- Limited access to site can complicate construction
- May pass erosion problems downstream

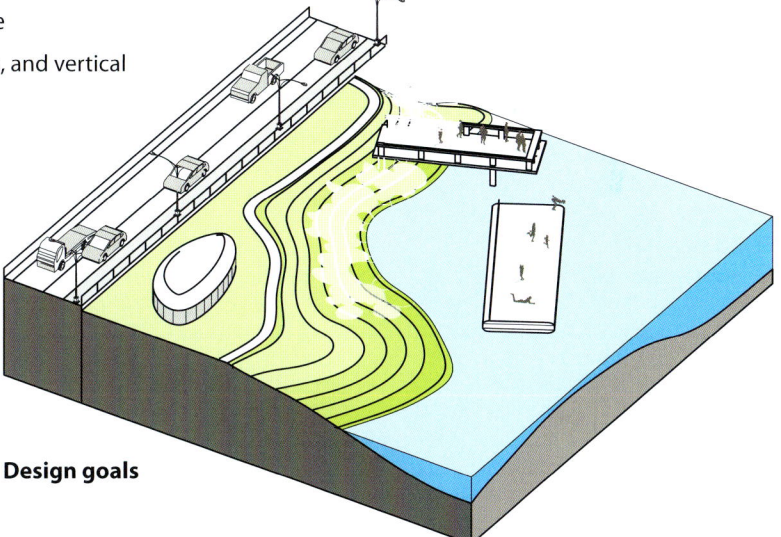

DESIGN GOALS
- Improve access to water
- Provide connections between the shore and the top level of the revetment
- Incorporate vegetation into hard surface
- Establish network of promenades, paths, and vertical circulations

Design goals

PART II: HARD-PROTECT STRATEGIES

Steps to water

Steps can provide access to the water while protecting against soil erosion.

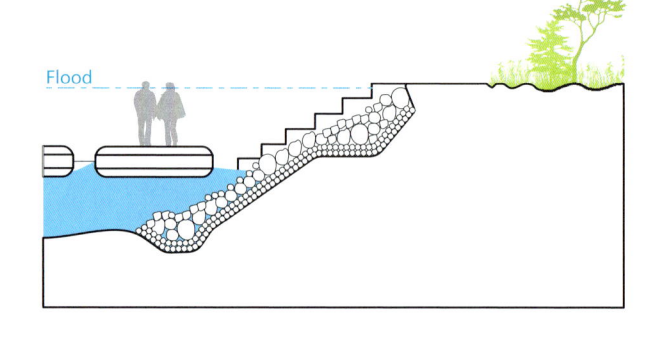

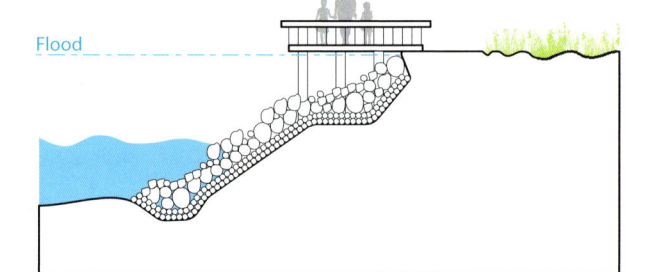

Elevated platform

On top of the revetment, additional structures can be extended to provide views to the water.

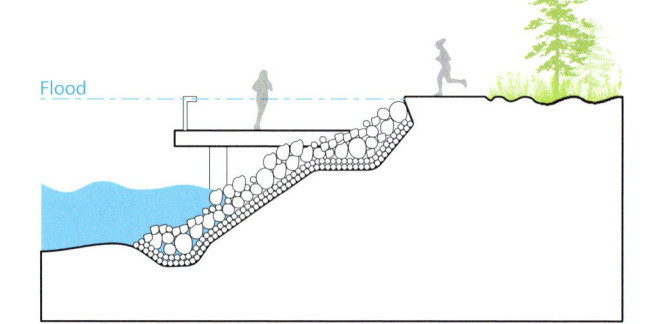

Floodable trail

Under lower risk conditions, designing floodable trails can stimulate the recreational use of flood zones.

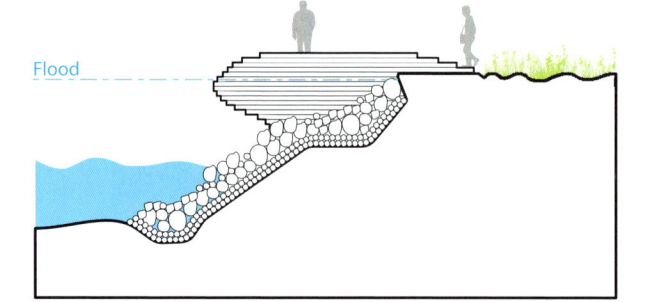

Viewing deck

Viewing decks can be integrated within the revetment.

CLEVELEYS COASTAL PROTECTION, CLEVELEYS, UK

The Cleveleys Coastal Defense project is an example of how a design strategy for coastal engineering can lead to a beautiful new waterfront. The project began as an enhancement of the waterfront along the western coast of the United Kingdom to provide usable waterfront access while safeguarding the city against tidal flooding, storm surge, and sea level rise from the Irish Sea. The Cleveleys Coastal Defense project is a part of the larger Fylde Peninsula Coastal Programme that includes the protection of two additional towns—Rossall and Anchorsholme. The projects combine desirable public spaces with flood protection, including taking sea level rise into design standards. The sinusoid shape creates amphitheater-like viewing spaces, while the continuous steps accentuate the shapes and enable visitors to easily access the beach.

Cleveleys coastal protection, UK
Steps, amphitheaters, and distinctive lighting animate the revetment at Cleveleys.
(Source: mesmoland, Flickr)

6.4 BREAKWATER

A breakwater is a structure that forms a harbor and basin to protect the shore from the effects of waves, as well as to provide a safe place for fishing vessels to berth. Rock and concrete make up about 95 percent or more of all the constructed breakwaters.[5] There are two primary types of breakwater—fixed and floating.

The primary function of a breakwater is to absorb wave energy and therefore reduce the force of waves before they hit the shore. Breakwaters can act as a first line of defense to calm the water, reduce wave height, and prevent shoreline erosion. Although breakwaters do not completely stop waves, their ability to reduce shore impact can reduce flooding and protect people, buildings, and ecological systems. But breakwaters can be expensive to build and may not provide long-term benefits. In addition, there are ecological concerns with breakwaters, because they can change migration and habitats for local species.

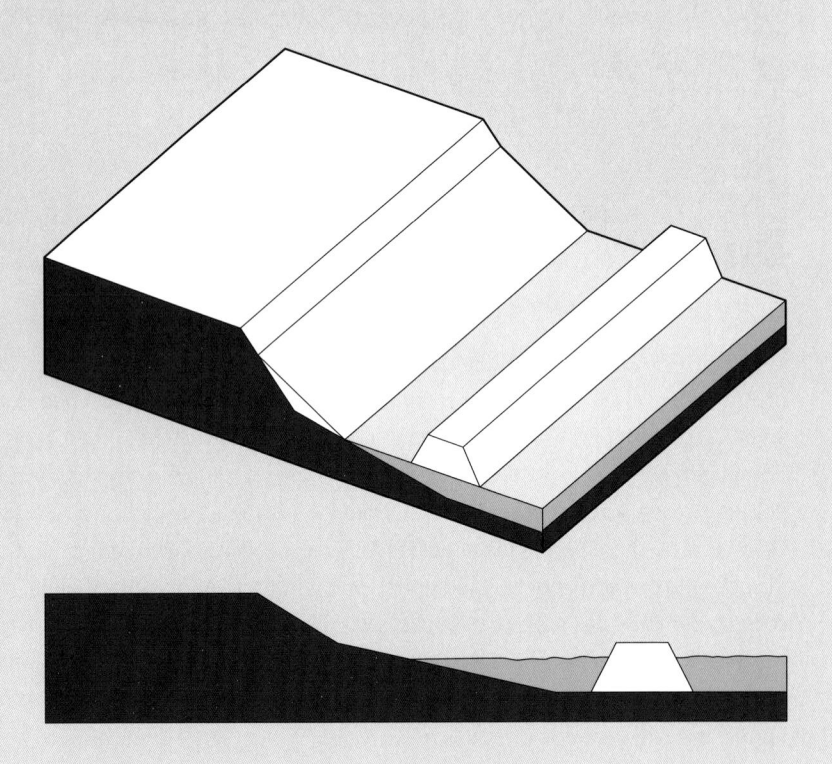

PERFORMANCE Breakwaters reduce wave force and shoreline erosion. Additionally, breakwaters contribute to reduced total flood levels during storm surge. In deep water, breakwaters become expensive—often too expensive for widespread usage. Floating breakwaters require anchors and support piles, which need a strong soil foundation. Floating breakwaters are used in places where fixed breakwaters may be submerged and therefore become ineffective, such as in areas of significant tidal fluctuations. Generally speaking, fixed breakwaters are better able to address the major wave forces of the oceanfront.[6]

PROS
- Reduce erosion
- Lessen wave impacts
- Create habitats
- Encourage recreational fisheries

CONS
- Must be supplemented by other strategies to keep out floodwater effectively
- More expensive in deeper areas, especially for water depths of more than 9 meters (30 ft)

DESIGN GOALS
- Combine infrastructure with other functions, e.g., art exhibition, water sports launch, aquaculture, and viewing decks
- Improve access to water
- Connect different levels
- Promote biodiversity through rock surface and pattern

Design goals

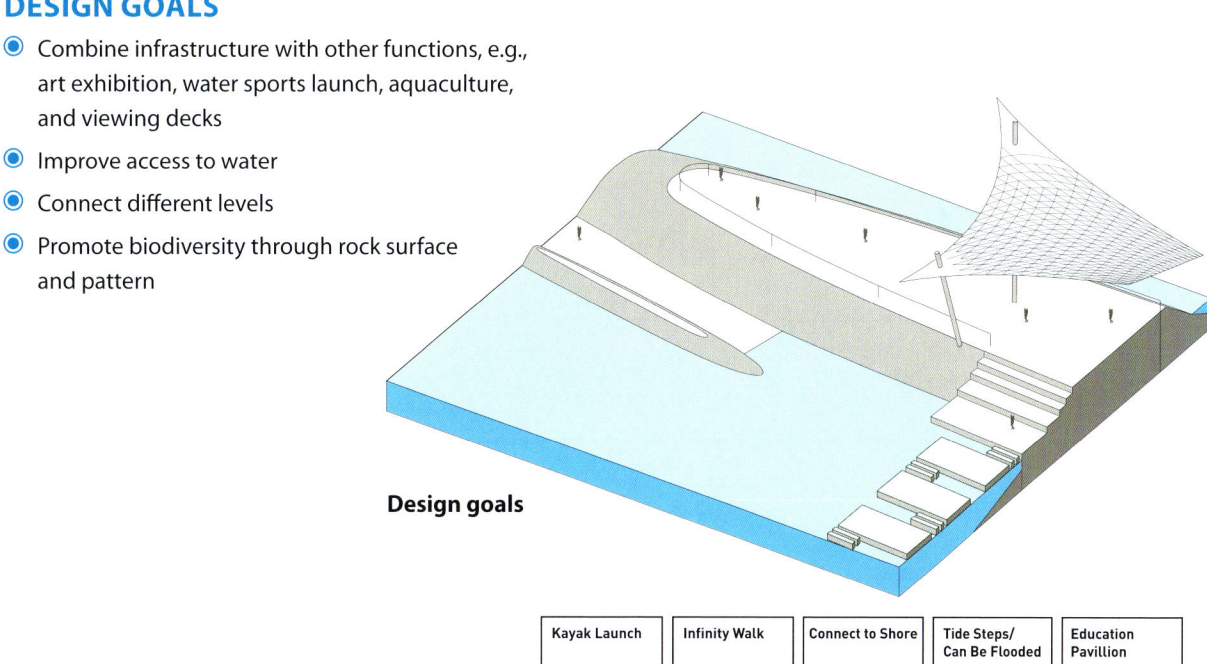

| Kayak Launch | Infinity Walk | Connect to Shore | Tide Steps/ Can Be Flooded | Education Pavillion |

PART II: HARD-PROTECT STRATEGIES | 67

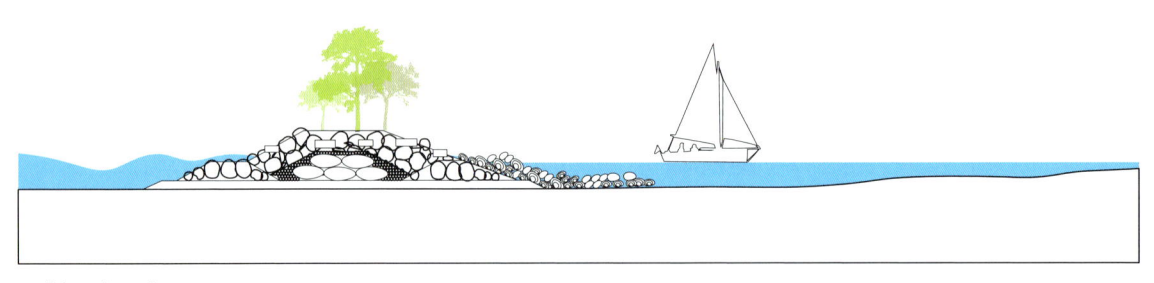

Habitat breakwater

Permeable breakwaters are more effective in connecting inner- and outer-bay ecosystems.

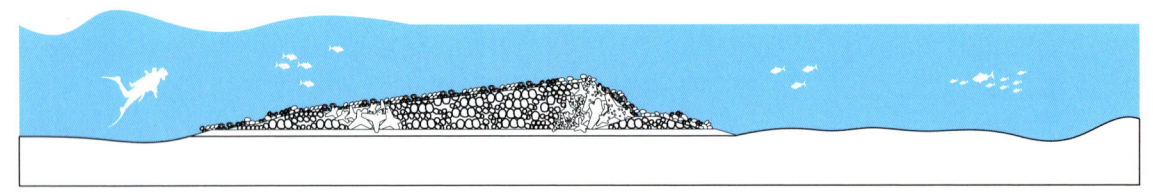

Constructed reef

Submerged breakwaters can become a habitat for sea life.

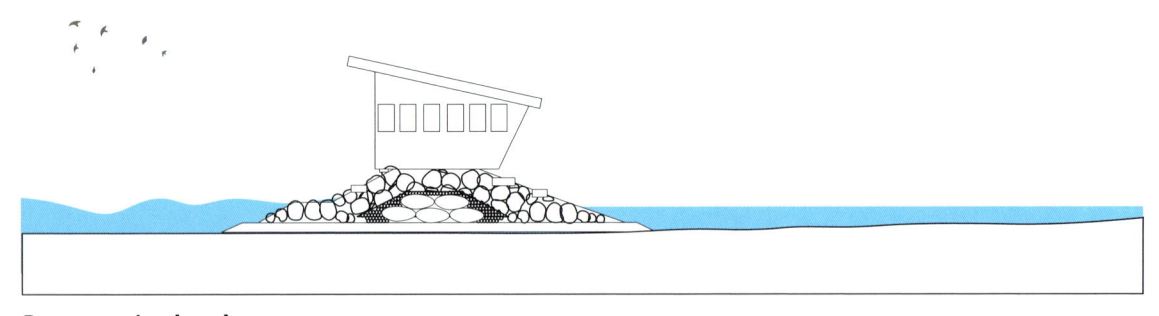

Programming breakwater

Civic programs, art exhibitions, or performances can provide people a reason to visit the breakwater.

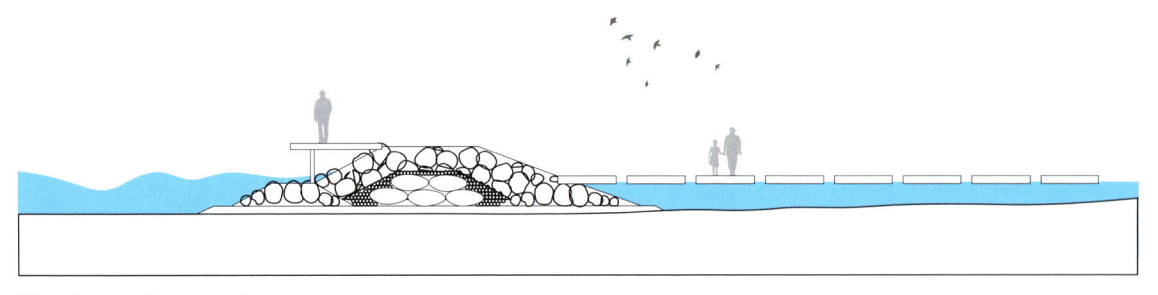

Floating path network

Improve connections between remotely located breakwaters and the shore.

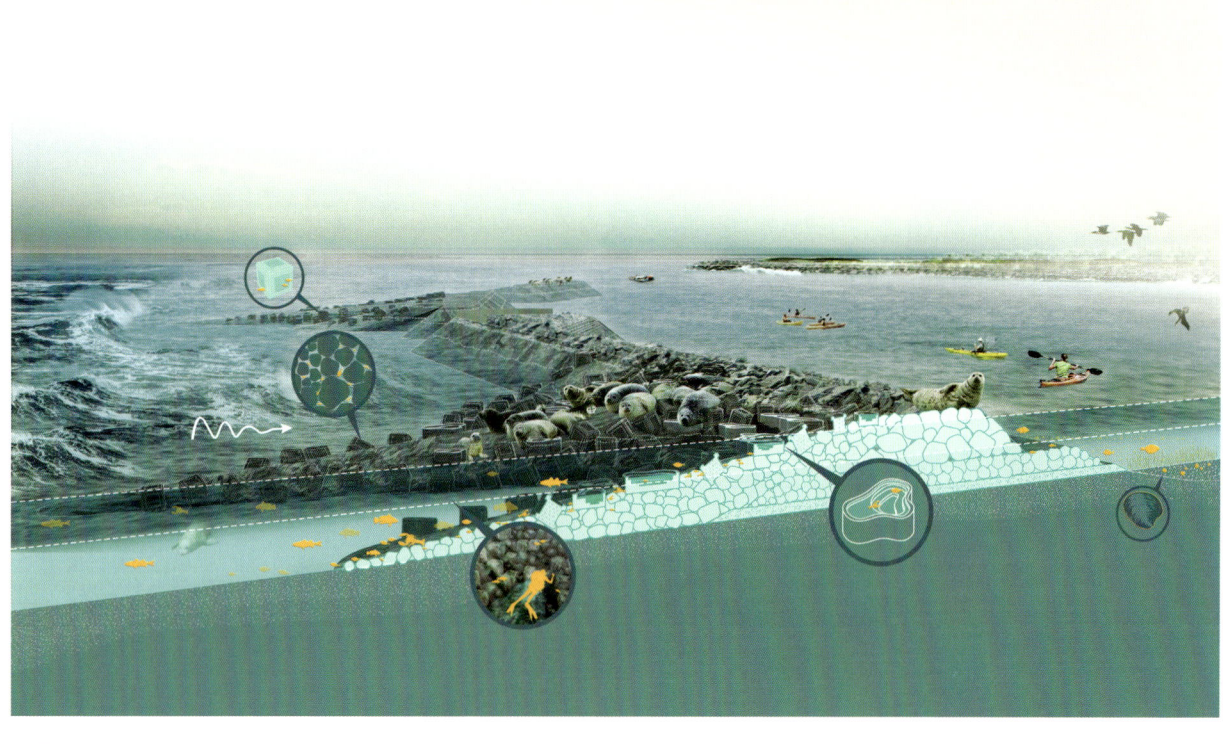

SCAPE—LIVING BREAKWATERS, NEW YORK CITY, USA

The Living Breakwaters project, located along the south shore of Staten Island and slated for construction in 2020, aims to reduce risk and increase environmental understanding among the local population, while also growing an ecological habitat and building social resiliency. The breakwaters are constructed as a rock mound covered with a bedding stone layer to protect against scour. The project proposes to revive ecologies, in particular through what SCAPE calls "reef ridges" and "reef streets," an outer layer of bioenhancing stones with a special concrete mixture that creates rocky protrusions, and a texture of narrow spaces, all of which provide opportunities for fish and shellfish species, such as oysters. The project also includes a water hub: an on-land facility to educate residents about the project and from which to organize beach cleanup events and birding walks.[7]

Living Breakwaters

Living Breakwaters reduce destructive wave energy and promote the growth of marine life such as oysters thanks to the creation of habitats in narrow spaces within rocks and encrusted surfaces.

(Source: the SCAPE team)

6.5 FLOODWALL

Floodwalls are vertical artificial barriers, either temporary or permanent, designed to withstand the waters from a river, waterway, or ocean. They are typically built of concrete or masonry, but glass versions exist as well. Floodwalls are a desirable form of flood protection because they can be located in open spaces or built into the urban fabric. Floodwalls can also serve other purposes, such as noise shields or visual barriers. Many floodwalls have recently been built to protect the coastline from flooding while also maintaining views and access to the waterfront, either through deployable, temporary barriers or through thoughtful permanent floodwall design, for instance, by raising the ground level on the inland side.

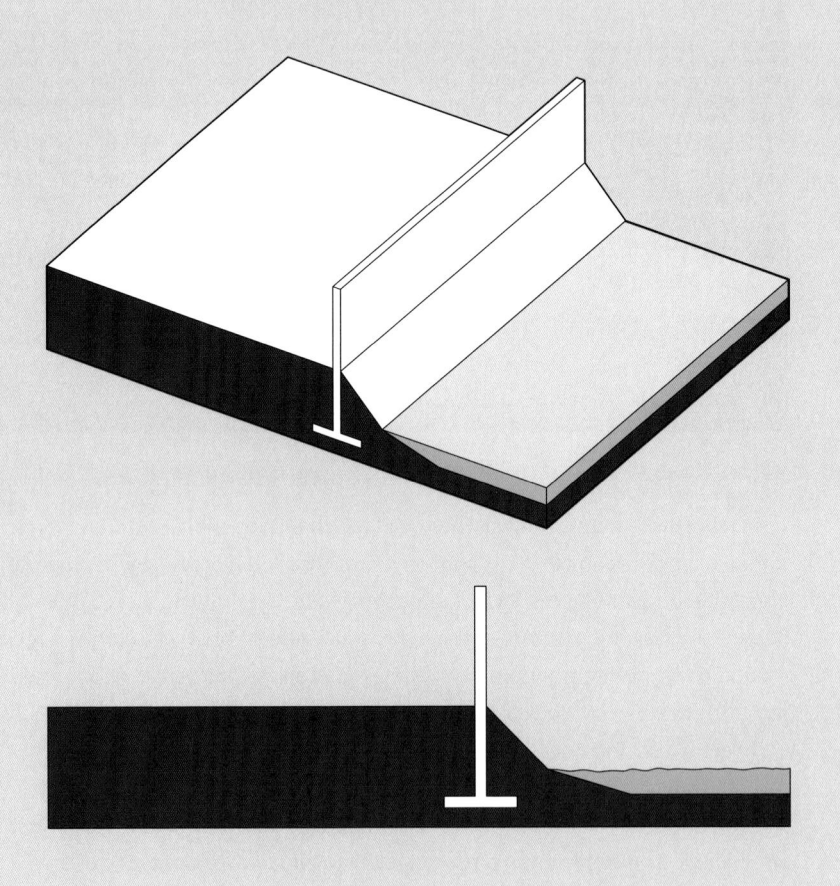

PERFORMANCE Floodwalls must be reinforced both above- and belowground to remain stable and successfully provide flood protection. When there is no flood, rainwater must be diverted or pumped to avoid a "bathtub effect" with water pooling on the upland side of the floodwall. Floodwalls may crack and leak if not built deep enough or high enough (potential for overtopping). Similar to seawalls, this hard solution can provide long-term protection but risks environmental and ecological harm to the surrounding area.

PROS
- Vertical flood protection requires a minimum of space and is suitable for densely populated areas
- The high top of the wall can be exploited to create a viewpoint
- Glass walls can serve as permanent flood protection without blocking the view of the bay/waterfront

CONS
- Reduction in a watercourse's natural retention space, which in turn increases the flood discharge peak and the danger of flooding downstream
- Intervention in floodplain dynamics, which will constrict the ecologically valuable space shaped by hydraulic fluctuations

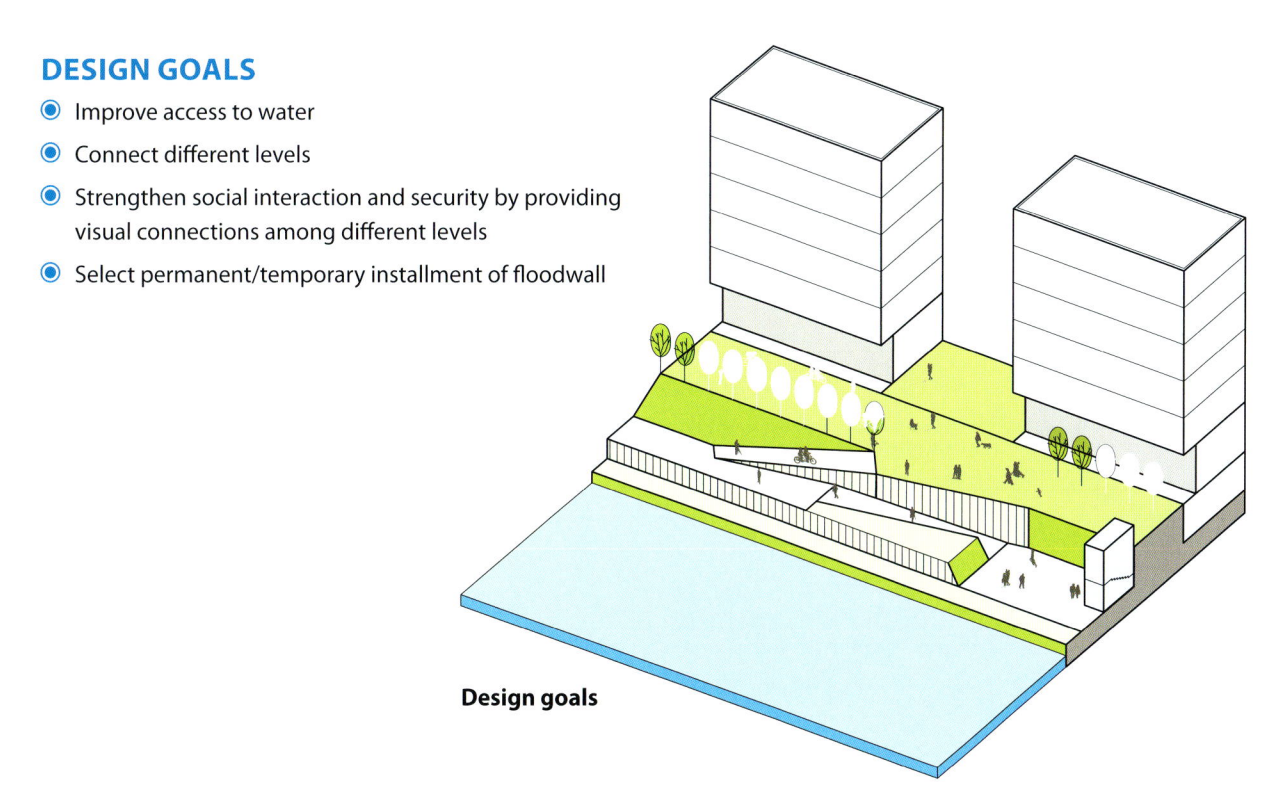

DESIGN GOALS
- Improve access to water
- Connect different levels
- Strengthen social interaction and security by providing visual connections among different levels
- Select permanent/temporary installment of floodwall

Design goals

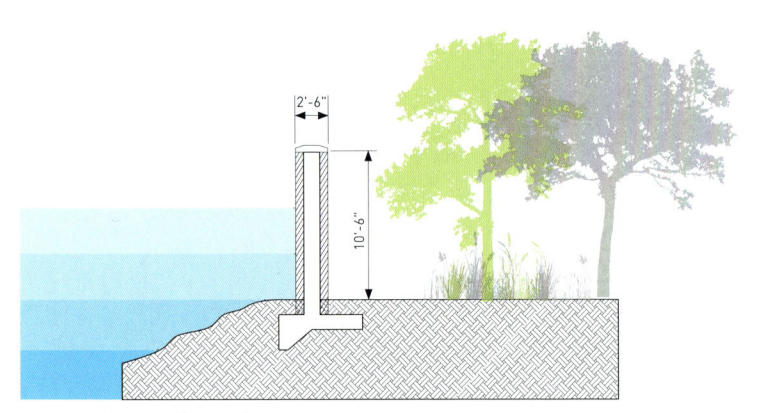

"T" wall example section

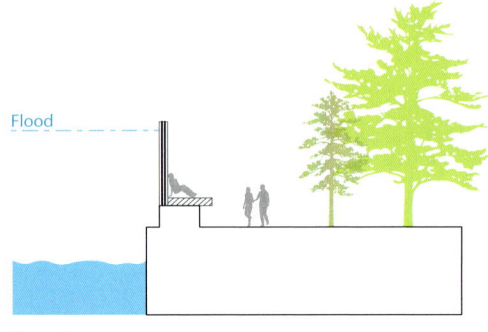

Bench
Activate engineered edge by incorporating sitting areas.

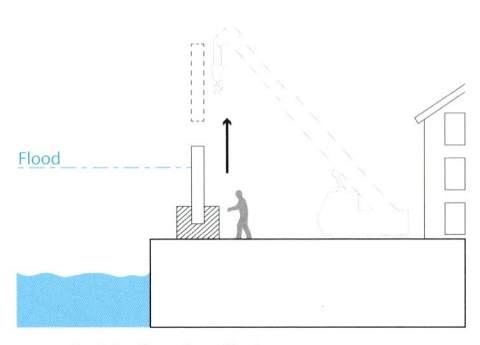

Attachable floodwall elements
When a permanent floodwall is not viable for its visual intrusion or other factors, demountable floodwalls allow for a solution, provided there is sufficient time to erect them.

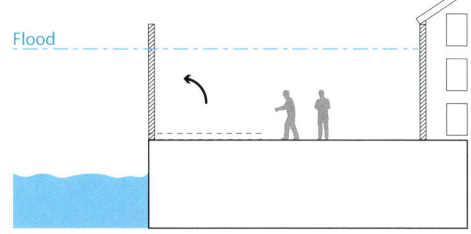

Fold-out floodwalls
Movable floodwall structures can be unfolded when needed.

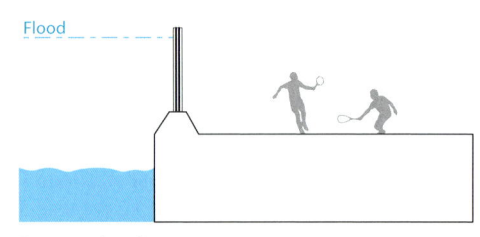

Recreation/sports
Activate engineered edge by accommodating recreational spaces.

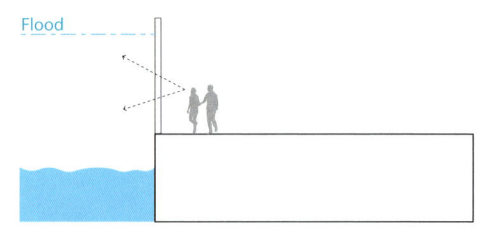

Transparent floodwall
A floodwall can be transparent to maintain a view to the water.

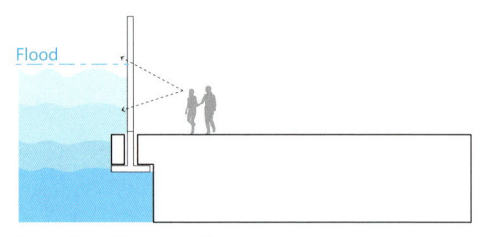

Floating floodwalls
Floating floodwalls can be automatically elevated by a buoy in the bottom of the wall.

MOBILE FLOODWALL, GREIN, AUSTRIA

Grein, in northeast Austria, is prone to flooding from the Danube River. To help protect the city from nuisance and heavy flooding, and at the same time avoid the visual nuisance of a permanent wall, Grein built a "mobile floodwall" in 2010. The floodwall is part of the Machland Dam superstructure, which regulates the river flow. The mobile floodwall is a vertical cantilever design with a permanent solid foundation and removable barriers, designed to hold back 4.6 meters (15 ft) of water. During a heavy rain event in 2013, the wall successfully held back as much as 4.3 meters (14 ft) of floodwater.

Mobile floodwall, Grein

Grein's mobile floodwall holds back up to 15 feet (4.6 m) of water from the Danube River.

(Source: Mitsuhiko, Reddit)

6.6 DIKE

Dikes, also commonly known as levees in the United States, are embankments made of artificial or natural materials, including earth. Dikes are designed to protect against floodwaters and can be multipurpose (see section 6.7). Some dikes, especially in the Netherlands, have roads or pedestrian crossings running along the top. Other dikes have floodwalls on top to increase the height of surge protection. Dikes are a fairly common form of flood protection and can be seen in Europe and North America.

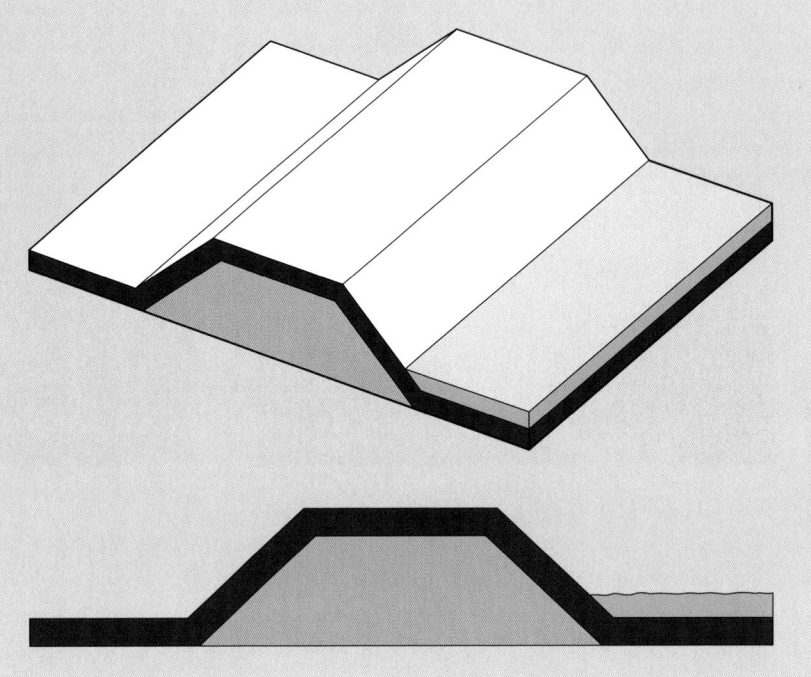

PERFORMANCE Dikes protect from storm surge but not from wave forces, so are more suitable for low-lying areas than waterfronts such as rivers, lakes, or polders. Dikes may encourage urban development in a flood-prone area. But dikes can fail because of overtopping, erosion, and slides within the foundation soils or embankment. Dike failure can be particularly damaging when dense urban development occurs immediately behind it, as Hurricane Katrina demonstrated in New Orleans.[8] The height of dikes provides opportunities for viewing points, and a network of dikes provides an elevated emergency route system during times of floods.

PROS

- A dike can be multipurpose (see section 6.7); for instance, as long as it is not submerged for most of the year, it can be developed into a park
- The height of dikes allows for various waterfront views for pedestrians
- The sloping features of a dike allow for many design ideas for recreational uses, including slides and waterfall steps

CONS

- Artificial levees can lead to an elevation of the river bed, especially at alluvial rivers with a lot of sediment
- Wave overtopping can cause dike failure

DESIGN GOALS

- Connect different levels with vertical circulation, ramps, stairs, and steps
- Activate the edge with paths, viewing decks, and other services
- Design the embankment

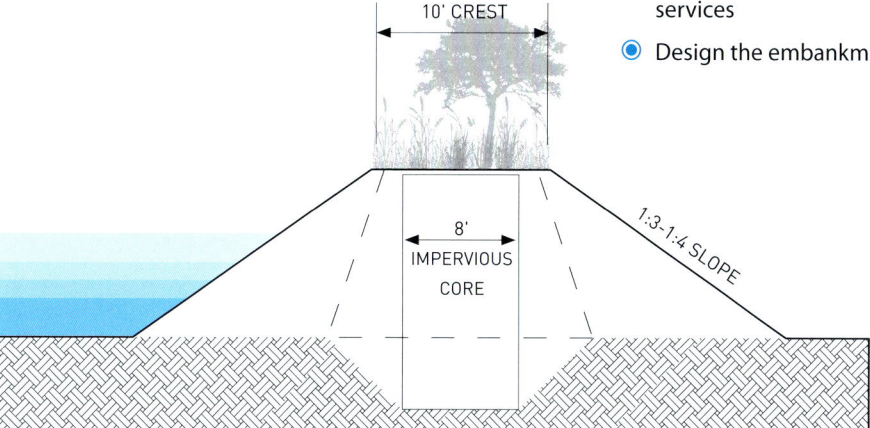

Design goals

PART II: HARD-PROTECT STRATEGIES | 75

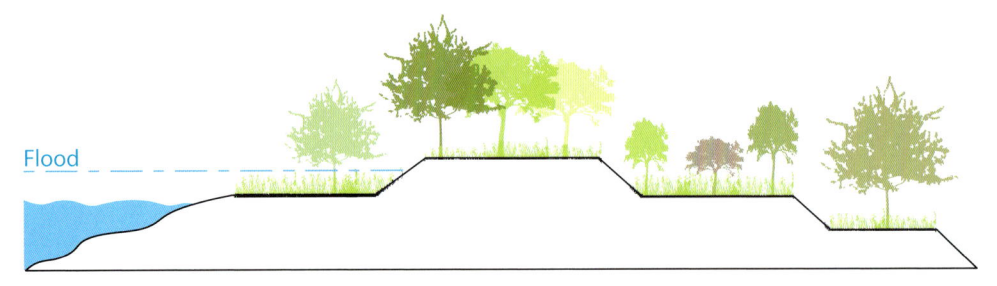

Park The difference in elevation can be resolved by creating a sloped park over the proposed dike.

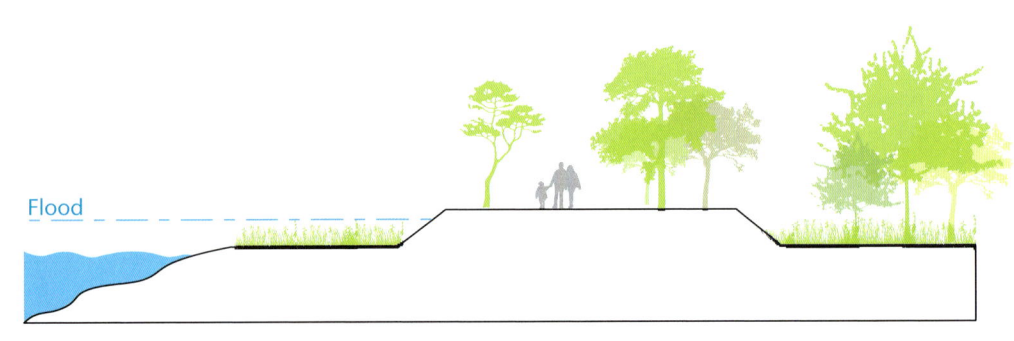

Path and trail The park can be connected with a network of paths and trails.

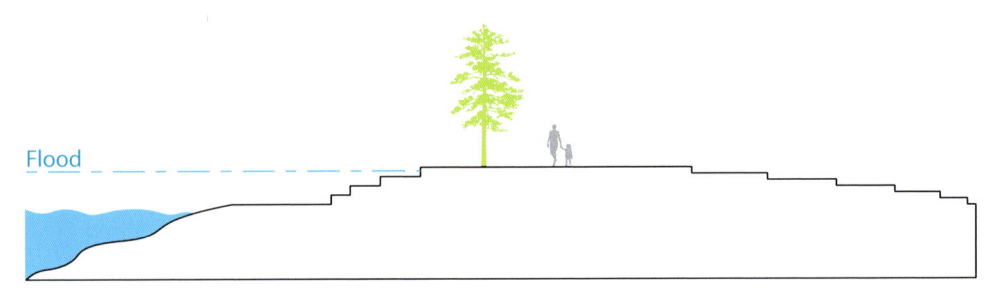

Steps Hard surfaces, such as steps and ramps, can be used to overcome the elevational difference.

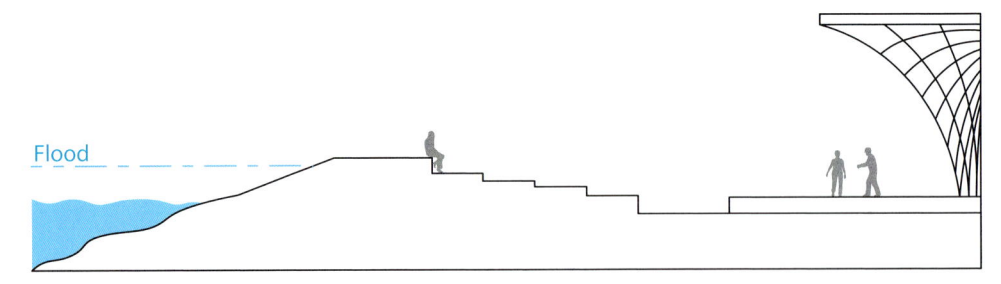

Amphitheater The dike's slope can be used as an auditorium for performances and to activate the edge of the dike.

MOTORWAY DIKE, THE NETHERLANDS

The Motorway Dike (otherwise known as Afsluitdijk), in the north of the Netherlands, was constructed between 1927 and 1932 and is part of the Zuiderzee Works. It is 32 kilometers (20 mi) long, 90 meters (295 ft) wide, and rises 7.25 meters (24 ft) above sea level. The dike was constructed to dam off the Zuiderzee, create the new freshwater lake of the Ijsselmeer, and seal it off from damaging salt water of the North Sea. The dike successfully created a freshwater lake, protected valuable farmland, and serves as a vital transportation network along the Dutch coast. The Motorway Dike has shipping locks and discharge sluices on both sides, allowing for access in and out of Ijsselmeer, if needed. The American Society of Civil Engineers declared the Afsluitdijk, as well as the Delta Works along the south shore of the Netherlands, as among the seven wonders of the modern world.[9]

Motorway dike

The Afsluitdijk (Motorway Dike) in the Netherlands is integrated with a highway.

(Source: jbdodane, Flickr)

6.7 MULTIPURPOSE DIKE

A multipurpose dike, also known as a multipurpose levee, functions very similarly to a regular dike (see section 6.6) but provides additional services. For example, a multipurpose levee can provide space for additional infrastructure such as roadways, or host residential and commercial space, storage, or open space. Dense urban centers, such as New York City, are highly suitable for multipurpose levees, because these structures can provide flood protection and add parks and residential and commercial space to the city.[10]

PERFORMANCE Like regular dikes, multipurpose dikes provide protection from storm surge events. Dikes are more effective against heavy surge if built wider, for instance, the Dutch "staircase" dikes, or when combined with armored revetments. Multipurpose dikes are found to be less suitable for ocean fronts, where a seawall will better attenuate wave force. Multipurpose dikes require a relatively large amount of land in comparison to other hard flood protection measures.

PROS
- Possible integration into the urban fabric, roads, and buildings
- Creation of a lower risk evacuation waterfront on higher ground
- Possible integration with natural strategies of coastal protection, including sand dunes

CONS
- Integration with buildings can increase risk when levees fail

DESIGN GOALS
- Overcome elevation differences with ramp, stairs, and grading
- Program the edge with paths, viewing decks, and parks
- Resolve highway traffic with pedestrian crossings
- Incorporate traffic on top of or below dike

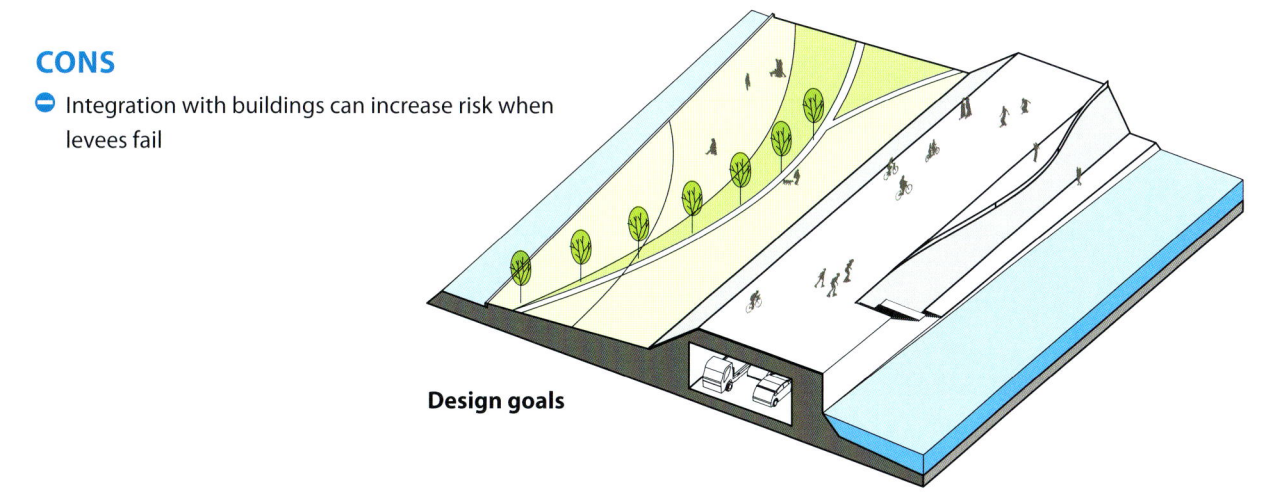

Design goals

Typical multipurpose section

A Rotterdam invention is the "staircase" dike: multiple flat platforms or "steps" that allow for multifunctional use of dikes, instead of a typical dike's sloped surface that does not allow for human activity. The increased width of the dike by the different platforms simultaneously strengthens the dike.

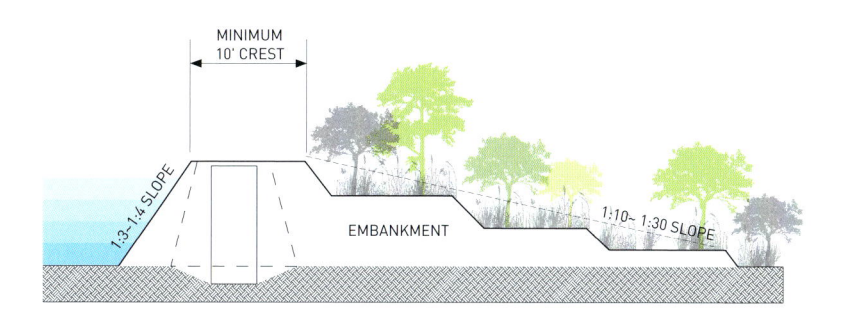

MINIMUM 10' CREST
1:3–1:4 SLOPE
EMBANKMENT
1:10~1:30 SLOPE

PART II: HARD-PROTECT STRATEGIES | 79

Highway on top of or underneath the dike

Incorporate highway into dike.

Platform for high-density development

Within dense urban areas, the dike can provide protection from storms and serve as space for underground parking.

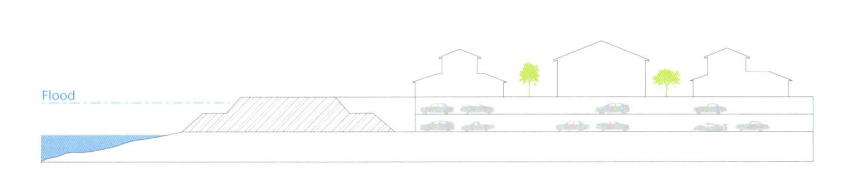

Greater park system along dike

Combined with open space, dikes can create a greater park system for a city.

Dakpark, Rotterdam, The Netherlands

A dike becomes a multipurpose dike at Dakpark, complete with a park, retail, and a parking garage.
(Source: Sant en Co Landschapsarchitectuur)

Dakpark water feature

A child climbs up the dike while water cascades down the steps.
(Source: Sant en Co Landschapsarchitectuur)

DAKPARK, ROTTERDAM, THE NETHERLANDS

The Dakpark, Dutch for "roof park," is a building, a park, and a dike. The project resulted from a design competition for an area with disused rail tracks. The Dakpark is Rotterdam's first staircase dike, with multiple flat steps, instead of the typical sloped surface, to allow for human activity as well as to strengthen the dike. On one side of the multipurpose dike is a brick façade with storefronts. The other side is covered by a park, which makes the parking garage, retail building, and service road disappear. The park includes various sections, including areas for children and their parents, and offers new views of the city.

Parking garage at Katwijk aan Zee, The Netherlands

Disguised as a dune, this multipurpose levee includes a parking garage.

[Photo credit to come from author.]

Entrance to parking garage, Katwijk aan Zee

A car drives down into the parking garage within the multipurpose levee.

[Photo credit to come from author.]

PARKING GARAGE, KATWIJK AAN ZEE, THE NETHERLANDS

The parking garage at Katwijk aan Zee is one component of a multipurpose levee located on the mid-Netherlands coast. The levee at Katwijk aan Zee protects the resort town from coastal flooding while including a below-grade parking garage, with entrances and exits located along the levee. From the outside, the levee looks like dunes with local grass plantings atop, which blend nicely with the surrounding landscape.

6.8 SURGE BARRIER

Surge barriers, fixed dam structures with movable gates, provide some of the highest levels of protection from coastal storm surge. Surge barriers protect best when coupled with protection measures such as shoreline levees, seawalls, and/or pumps. During dry conditions, a surge barrier's gates will remain open to allow the free flow of water, and vessels. However, prior to a storm, the gates will be closed.

There are several different types of surge barriers. These include sector gates, vertical lifting gates, and, on a smaller scale, tide gates. Surge barriers can vary in design, and also have several associated structures whose designs can vary, including pumping stations to pump water away from the barrier, navigational locks, and adjacent levees or seawalls. Surge barriers, whether used for large waterways or on small streams, require extensive maintenance and monitoring.[11]

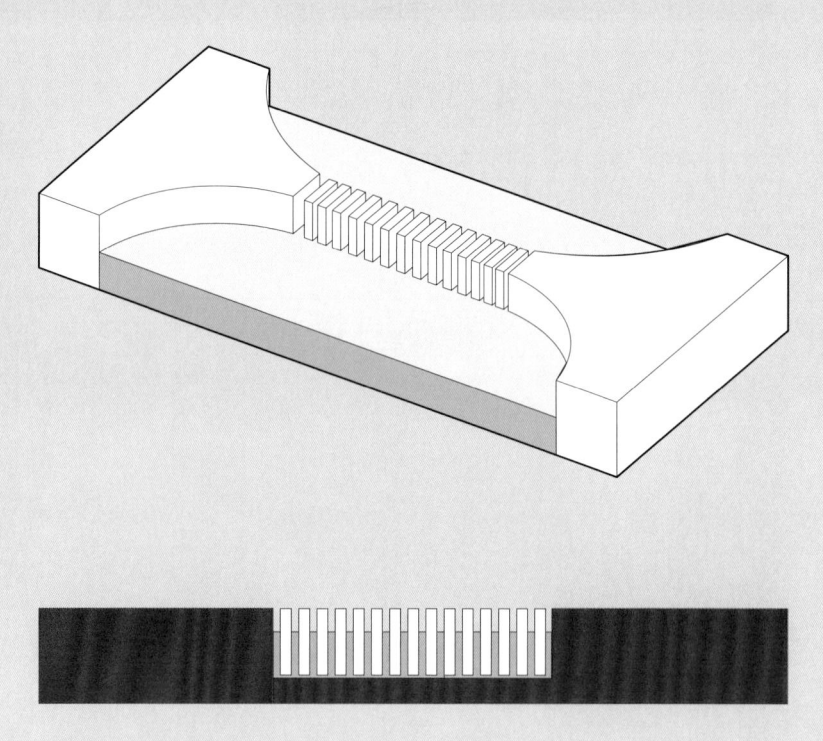

PERFORMANCE Surge barriers are often the most appropriate form of flood protection for navigable channels and waterways. Because surge barriers require connections to the shoreline on both sides of the barrier, including higher elevations on the shoreline, they are found to be most feasible in water bodies that can be closed off, such as inlets. Depending on the site's geography and the construction of other forms of flood protection in the area, opportunities for larger surge barriers may exist.

PROS
- Can be controlled manually with sector gates or immovable barriers
- Can be very effective at protecting area inside barrier if engineered correctly

DESIGN GOALS
- Develop flood management strategies for area outside the barrier
- Incorporate places for human interaction with engineered infrastructure
- Program the edge with viewing decks
- Provide access to water
- Include educational programs such as museums

CONS
- Very high installation and operation and maintenance costs
- Human error poses a risk when barriers must be manually opened and closed
- Risks of flooding outside the barrier
- Risks of ecological and estuarine system changes from barriers

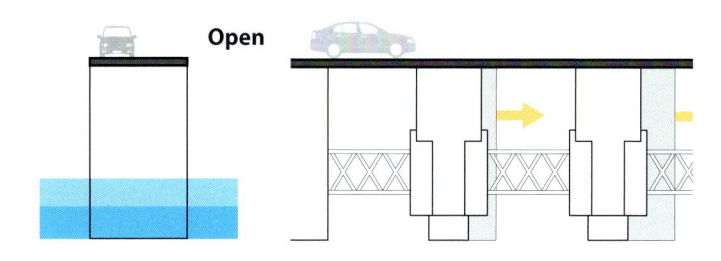

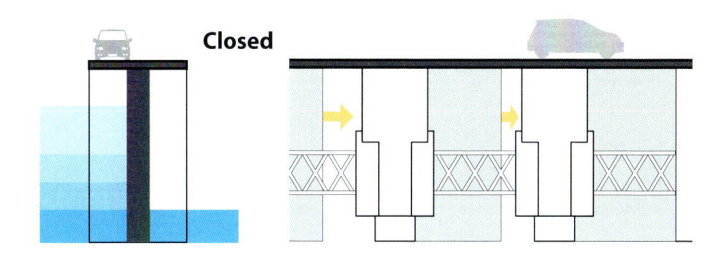

Open

Closed

PART II: HARD-PROTECT STRATEGIES | 83

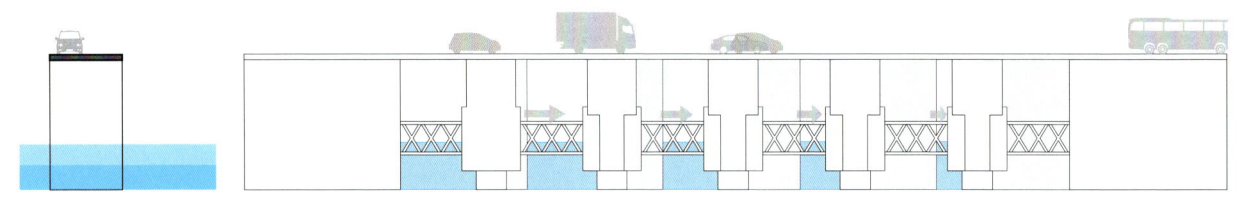

Surge barrier with vehicular right of way
The top of the barrier can provide vehicular connections across a bay or river.

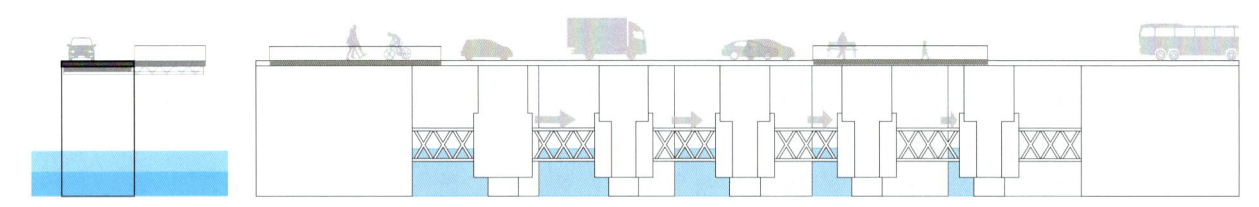

Surge barrier with viewing deck
Adding viewing decks provides emergency stops and sightseeing activities for tourists.

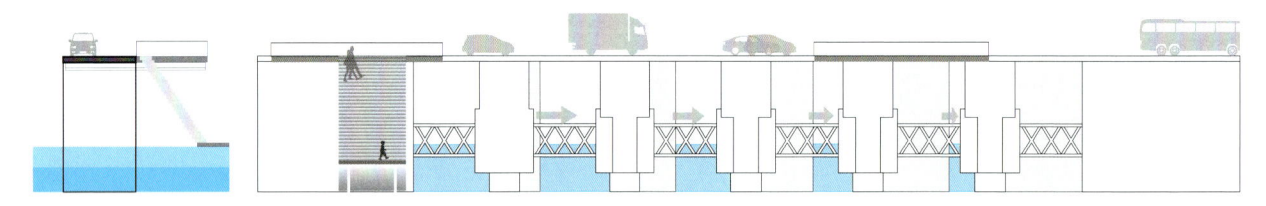

Surge barrier with access to water
Viewing decks along the barrier can provide opportunities for accessing water; however, this interaction needs to be coordinated with the operation of the barrier.

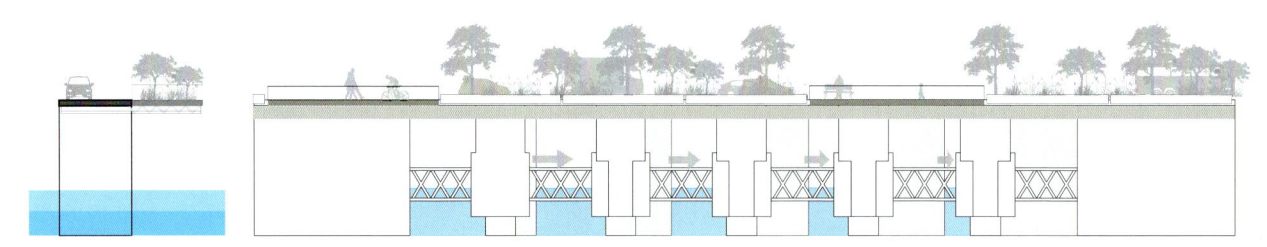

Surge barrier with pedestrian park
Parks can provide recreational amenities to barriers.

MARINA BAY BARRAGE,
SINGAPORE

Constructed between 2005 and 2008, Singapore's Marina Bay Barrage is an example of a storm surge barrier that can also provide other benefits besides flood management, including a large freshwater reservoir and art galleries. It created the city's largest and first urban reservoir, which has become the key feature of the new central business district. The Marina Reservoir allows for activities including kayaking and dragon boat racing. The barrage's pump house regulates the water level and doubles as a museum and education center with a green roof for recreation.

Marina Bay Barrage, Singapore
Singapore's Marina Bay Barrage created the city's first urban freshwater reservoir. The pump house (right) features a museum and green roof.
(Source: CEphoto, Uwe Arana, Wikimedia Commons)

THE OOSTERSCHELDEKERING,
THE NETHERLANDS

The Oosterscheldekering surge barrier was constructed to protect the Zeeland region of the Netherlands from flooding, particularly in the wake of the devastating 1953 North Sea flood. It connects the islands of Noord-Beveland and Schouwen-Duiveland with a highway. It was originally intended as a dam, but public protests demanding access to and from the North Sea led to a surge barrier with sluice gates. During dry weather, the gates remain open. In preparation for a storm, the gates will close. Keeping the surge barriers' gates open allows for the protection of local marine life, and the gates have been closed only twenty-five times since completion. The Oosterscheldekering is designed and built for a 200-year life and is considered a model for surge barriers around the world.

Oosterscheldekering, The Netherlands
The 9-kilometer-long (5.6-mi) Oosterscheldekering, a moveable barrier with huge sluice gates, took a decade to construct.
(Source: Zairon, Wikimedia Commons)

CHAPTER 7: SOFT-PROTECT STRATEGIES

▶ The second section on local solutions features the "soft" protect strategy, also known as a systems strategy, a way to hold the line of defense using nature-based systems, such as dunes. The examples featured in this section include living shorelines, dunes, and floating islands. The systems strategy is now often combined with hard-protect strategies, making solutions of hybrid gray + green—for instance, a "sandy foreshore" dike reinforcement.

Soft solutions take up more space than hard solutions and are vulnerable to human use—for instance, the dunes, held together by dune grass, which is tolerant to extreme glare and high salinity but not to trampling. However, a major advantage of soft solutions is their ability to provide habitat for species and promote biodiversity.

7.1. Living Shoreline

7.2. Dunes and Beach Nourishment

7.3. Floating Island

7.1 LIVING SHORELINE

Living shorelines are gently sloping natural banks that reduce shoreline erosion, protect coastal ecosystems, and help reduce storm surge strength along the coastline. Living shorelines use plants, sands, and limited hard landscape (hardscape) approaches such as concrete and rocks to ensure and maintain the natural habitat. They are increasing in popularity over traditional hard-protect strategies such as bulkheads. For instance, Brooklyn Bridge Park, one of New York's most recent waterfront parks, includes riprap shorelines. They proved to be quite resilient during Superstorm Sandy in 2012, with even the salt-tolerant vegetation surviving.[1]

PERFORMANCE Relative to hard-protect strategies such as bulkheads, living shorelines take up more space. They work best in low to moderate flooding areas and can be paired with levees to provide more substantial flood protection. Complementary flood protection measures include upland strategies such as waterfront parks and/or inland strategies such as groins and surge barriers. Living shorelines are generally low-cost and low-maintenance and provide modest wave attenuation for storm surges. However, living shorelines can alter shoreline ecology and erosion, and therefore require moderate maintenance to ensure a healthy ecosystem. Because they are covered in vegetative growth, they can double as parks. Whether embedded in wetlands, marshes, or mangroves, living shorelines can be exciting areas for recreation, accessed through elevated walkways and floating paths, and enhanced with education centers and tourist facilities.

PROS

- Can serve as fish and oyster habitats
- Can improve water quality because oysters are also filter feeders
- They are natural coastal buffers that help break up wave energy

CONS

- Shoreline erosion

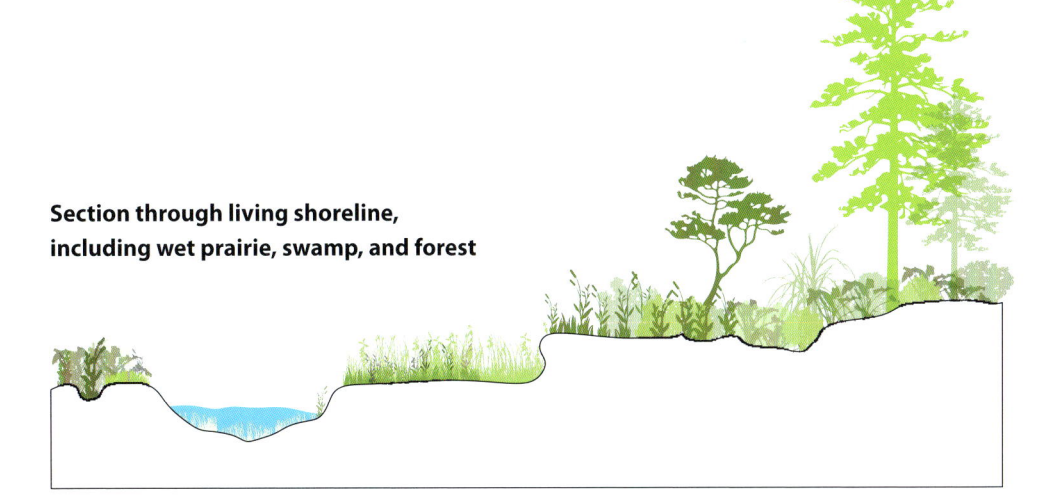

DESIGN GOALS

- Protect and preserve resilient natural system to handle sea level rise
- Minimize human disruption
- Weave paths through nature
- Incorporate educational program

Section through living shoreline, including wet prairie, swamp, and forest

Floating path

Separating paths from the living shoreline can minimize human interference and preserve the natural system.

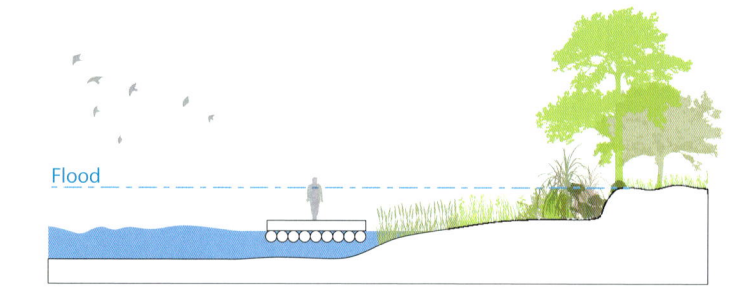

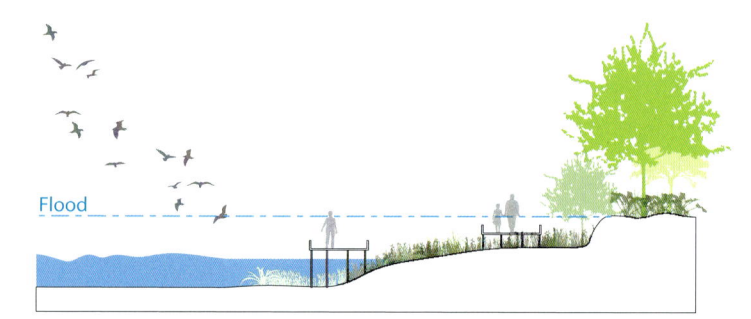

Elevated path

Paths can be elevated to allow closer observation for educational purposes.

Dedicated path

Strategically clearing paths at grade can enrich eye-level experience of the living shoreline.

Elevated structure

Any structure above the living shoreline should not disrupt the natural system and its ability to deal with storm surges.

STUMP SOUND SHORELINE, NORTH CAROLINA, USA

As estimated by scientists, North Carolina has lost between 50 and 80 percent of all its oyster reefs since 1900 due to overfishing, disease, and deterioration of habitats. The North Carolina Coastal Federation purchased the Stump Sound shoreline to save it from development. It created a living shoreline that is almost a laboratory for stabilizing shorelines, testing various natural approaches. The shoreline serves as a buffer for stormwater runoff and a lifesaver for the oyster population. Because living shorelines improve water quality, they can preserve the marshes that accommodate oysters and juvenile fish and help restore decaying oyster habitats.[2]

Stump Sound shoreline, North Carolina, USA

The Stump Sound shoreline features a path for visitors such as officials and property owners to learn more about oyster restoration and other living shoreline techniques.

(Source: North Carolina Coastal Federation)

7.2 DUNES AND BEACH NOURISHMENT

Dunes protect the coastline by their ability to dissipate waves and protect against storm surge, particularly because they are located at a higher elevation than the surrounding shoreline. To prevent erosion, they are often stabilized using natural grasses or fencing networks that help keep the sand in place and prevent it from washing away too quickly. However, this cannot prevent long-term erosion and movement of sand. Reinforced, or "armored," dunes are seawalls covered in sand that are more able to withstand heavy wave action.

Beach nourishment, which can go hand-in-hand with dunes, is a flood protection measure in and of itself. Stable beaches with adequate sand and fill help against storm surges, reducing wave strength as storms come ashore. Beach nourishment projects are considered "near-term" rather than "long-term" because erosion will eventually alter beach structure. The artificial shaping of beaches and dunes to reduce coastal storm impact is a common practice in Western Europe and the United States.

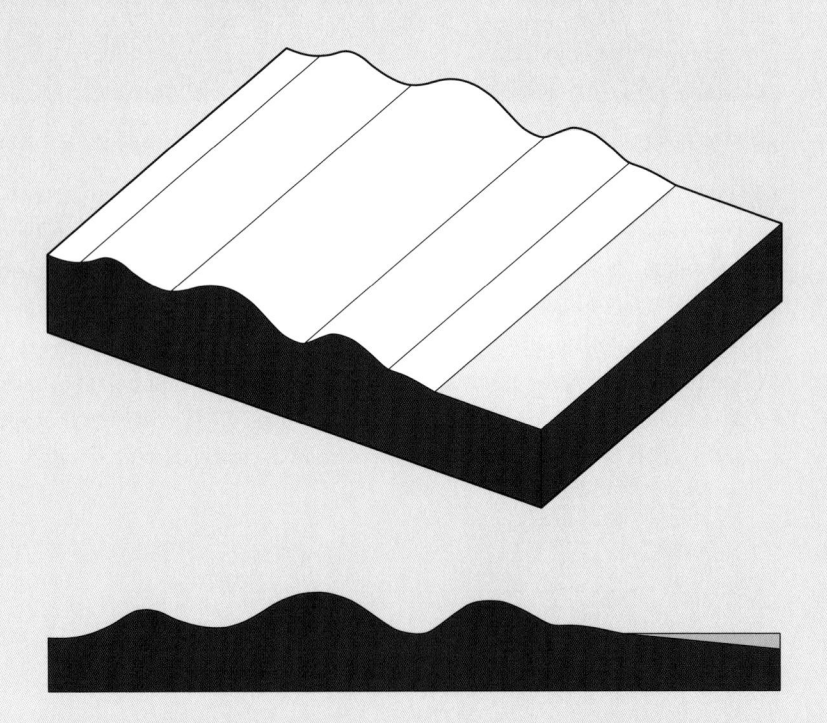

PERFORMANCE Although beaches and dunes can be a great alternative to seawalls, dikes, and other more imposing flood protection measures, they are noticeably less stable as compared to hard solutions. Although dunes are a more natural and visually appealing option than a seawall, they are prone to quick erosion during storms and periods of heavy flooding. Dune grass can help prevent dune erosion, but the grass is vulnerable to trampling. Dunes operate best when there is plentiful sand and sediment supply. They do require maintenance through beach nourishment projects, and have high rates of erosion, often making them costly to maintain.[3]

PROS
- Double dune systems allow for plant and animal habitats at the secondary dune, protected by the primary dune
- The lifespan of dunes can be extended by trapping sediment with grass, sand fences, and groins
- Relatively cost-effective as compared to hard solutions

CONS
- May erode faster than hard solutions
- Takes up more space than hard solutions
- Can be difficult to maintain

DESIGN GOALS
- Protect and preserve natural systems to handle storm surge
- Minimize human trampling on dune grass
- Strategically allow access to water through separate access paths over dune grass
- Minimize engineered infrastructure

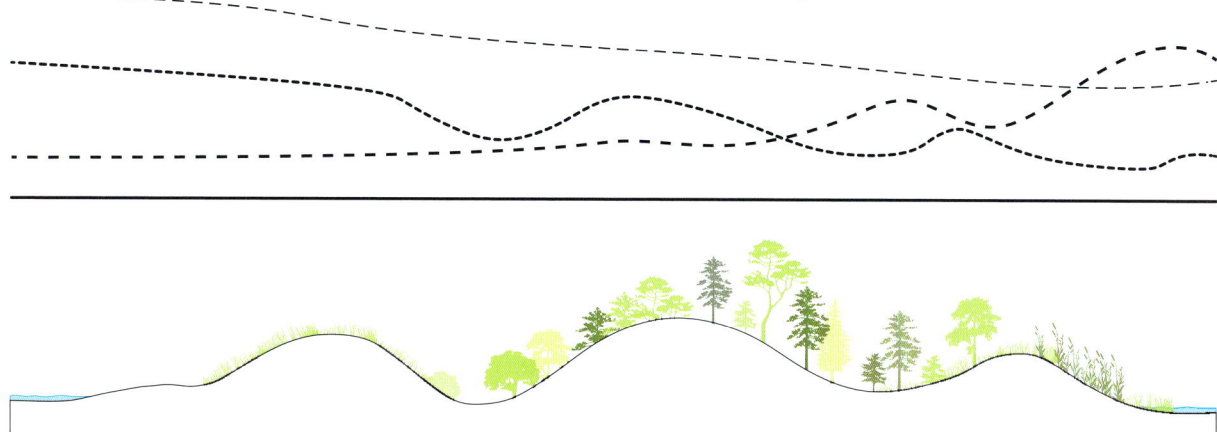

Section Dunes allow for intensive development on backshores, in contrast to troughs.
(Source: Adapted from Ian McHarg, *Design with Nature*, 1969

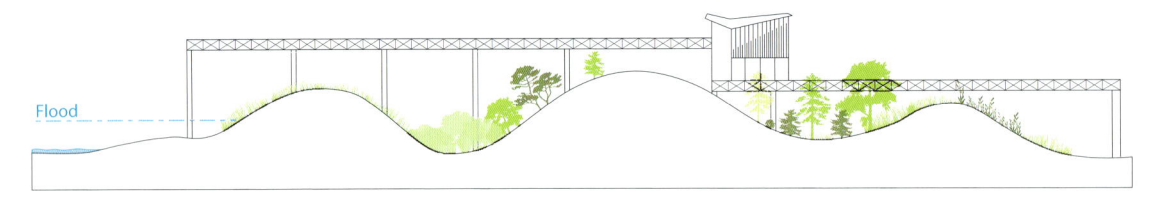

Elevated walkways
Separate pedestrian access is critical to preserving sensitive dune grass.

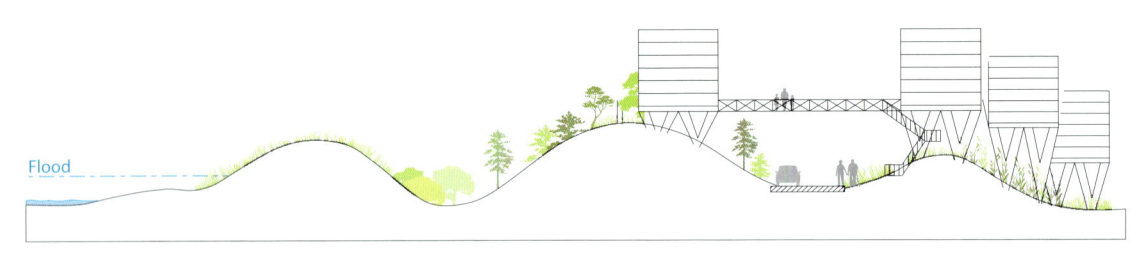

Retracted development
Development needs to be set back from the first two rows of sand dunes.

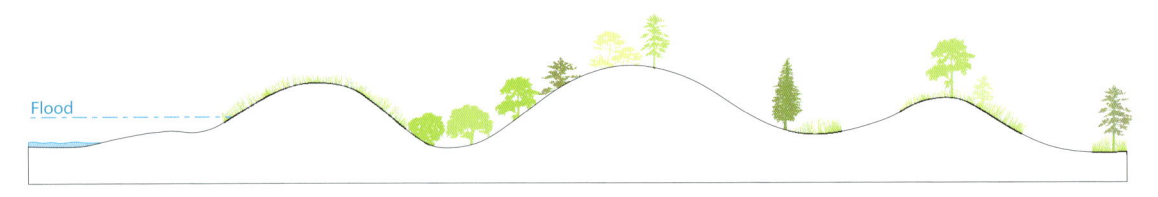

Preserve dune vegetation
Dunes should largely be restricted from human use and development, as these tend to destroy the natural systems that protect the dune from erosion.

SAND ENGINE, DELFLAND, THE NETHERLANDS

Usually beach nourishment occurs through the mechanic pumping and dumping of sand along the beach. In contrast, the Sand Engine, otherwise known as the Sand Motor, relies on natural forces to distribute the sand along the coastline, and possibly also inland. In 2011, a total of 21 million cubic meters (27 million yd³) of dredged material was pumped up by hopper vessels and placed into a hook shape protruding 1 km (0.6 mi) from the shore along the southern coast of the Netherlands.[4] As waves and currents move the sand along the coast, beaches are nourished, reducing wave action and erosion along shores. The Sand Engine is considered a cost-effective and natural way to protect coastal communities in the region and is a model for other flood protection schemes.

Sand Engine, Delfland, The Netherlands

The Sand Engine peninsula is a concentrated nourishment of sand that relies on wave forces to grow beaches for coastal safety and recreation.

(Source: Rijkswaterstaat/Joop van Houdt)

7.3 FLOATING ISLAND

Artificial floating islands are typically constructed of a thick, floating organic mat that can support plant growth. The islands can help dampen wave energy in sheltered water bodies—although this is as yet relatively untested—and environmentally remediate water. The upper portion of the mat is known as the root zone, and is made up of intertwining plant root. Under the root zone is the peat layer, which includes decomposed peat and decaying plant matter.[5] The peat layer can be very thick and is measured by the rooting depth of the plants. Underneath the peat layer is the water column with varying depths. A layer of organic sludge builds up under the water column, helping the floating island to remain afloat.

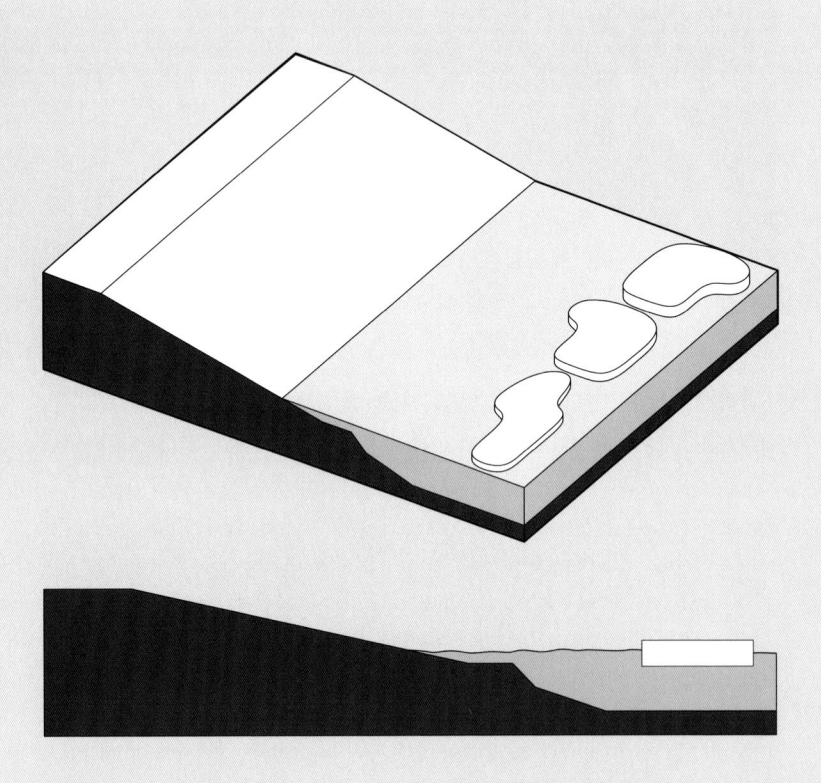

PERFORMANCE Floating islands, mats of organic matter that can help attenuate waves, act similarly to breakwaters to reduce shoreline erosion and protect the coastline from erosion. Floating islands work best for low to moderate wave forces, such as in sheltered bodies of water, as they generally cannot withstand strong wave forces.[6] They would need to be anchored into the seabed to do this. Mats of predominantly organic material tend to be more buoyant.

PROS
- Can be self-sustaining, and should not require much maintenance
- Inexpensive to install
- Can help improve water quality or provide biomass for use as animal feed or fertilizer

DESIGN GOALS
- Program islands to complement functions on the shore, such as recreation, residential, educational, and agricultural
- Improve access to land and other islands
- Allow height to be adjustable as sea levels rise

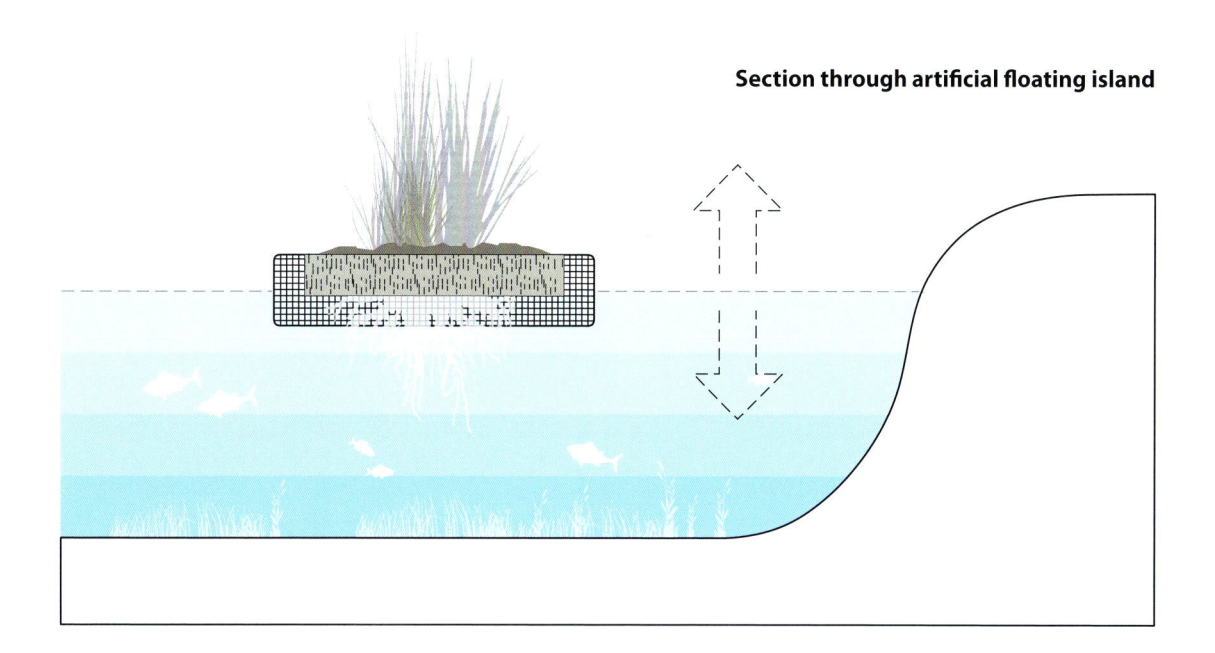

CONS
- Not suitable for most habitats and coastlines
- Cannot singularly prevent flooding; must be part of a larger flood protection scheme
- Expensive to install

Section through artificial floating island

Natural floating island

Floating islands commonly occur naturally in marshes and wetlands as a mass of aquatic plants, mud, and peat, and can be as large as several hectares.

Biofilters

Biofilters are a popular application of artificial floating islands and can produce biomass and environmentally remediate water.

Biofilter clusters

Biofilter clusters maximize the benefit of biofilters on a larger scale. Floating wetlands and reedbeds help manage and purify stormwater runoff.

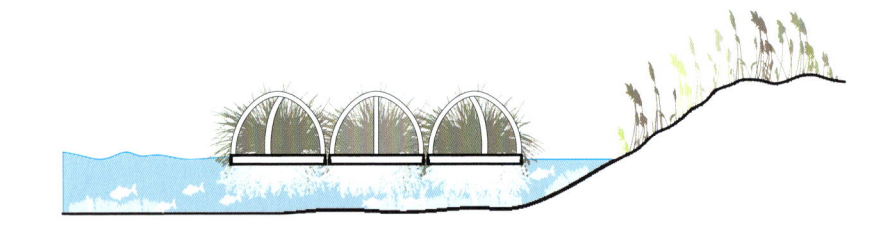

Recreation

Floating structures can be transformed into a recreational facility while maintaining their ecological functions.

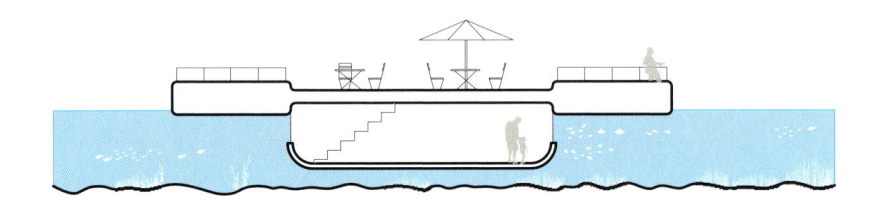

ISLE DE JEAN CHARLES FLOATING ISLAND PROJECT, LOUISIANA

Louisiana's Isle de Jean Charles in the Mississippi Delta is the epicenter of coastal land loss in the U.S. The island has lost 98 percent of its land due to rising sea levels and coastal erosion. In response, volunteers planted 187 five-by-eight-foot (1.5 m × 2.4 m) floating islands with up to sixty types of native plants. The floating habitat mattresses are made of recycled plastic bottles and filled with soil. The mats were then anchored along 1,500 feet of marsh. As the plants grow their roots into the water, they will trap sediment and prevent shoreline erosion.

Isle de Jean Charles Floating Island Project, Louisiana
The floating habitat mattresses help prevent shoreline erosion and create marine habitat.
(Source: Brett Milligan)

AQUA BIOFILTERS, CITY OF WAGGA WAGGA, AUSTRALIA

The Wollundry Lagoon floating wetlands and floating reedbeds are located in the central city area of Wagga Wagga, Australia. The city's lagoons, while highly valued natural resources, are polluted from urban runoff and have experienced fish kills and odor problems from stagnant water. The 120 square meters (1,292 ft^2) of floating wetlands within the lagoon provide enhanced water treatment for urban runoff, allowing for healthier lagoons. The floating wetlands have been established with a variety of native plantings, which are quickly colonizing the floating wetland biofilter media, allowing the wetland to mature and provide greater water treatment. During the establishment phase of the plants, bird netting prevents birds from consuming the young plants. Surprisingly, even trees have colonized the floating islands.

Aqua biofilters, City of Wagga Wagga, Australia
Reeds and other plants grow on a floating biofilter, which helps treat urban runoff.
(Source: Tom Duncan, AquaBiofilter™—www.floatingwetlands.org)

CHAPTER 8: STORE STRATEGIES

▶ In contrast to the "defend" strategies in the previous two sections, store strategies are a form of "accommodation." Instead of trying to keep water out, such a strategy aims to control and temporarily store water upland, to avoid flooding issues. Other than preventing the flooding of roads, buildings, and infrastructure, one of the key issues that store strategies help solve is the problem of backflow flooding. During intense storms with heavy rainfall, stormwater could flow back into the sewage system's discharge pipes, causing sewage in sewer lines to back up into houses, leading to damage and health hazards. Urbanization and the densification of cities compounds this problem, because it often leads to the increase of impervious surfaces, such as asphalt, which do not absorb water and further strain the drains.

This section features storm- and floodwater retention areas that are integrated into parks, polders, plazas, and streets. Channels and drains on rooftops and ground surfaces can flow stormwater to reservoirs, both underground and aboveground, where it is held and eventually discharged. Besides solving flooding issues, these solutions can make cities greener, while aboveground reservoirs enhance people's everyday experience with the seasonality and fluctuations of water. They show that the water cycle can be a planning and design principle for urban landscapes.

8.1. Floodable Plain
8.2. Polder
8.3. Floodable Square
8.4. Stormwater Infiltration

8.1 FLOODABLE PLAIN

Floodable plains are flat areas adjacent to a river or body of water that can be flooded when the water body's capacity is exceeded. Floodable plains use existing natural or urban environments to catch stormwater and control for floods. During nonflood times, the floodable plains can be dry and used for other purposes, such as recreation.

Although cities benefit from access to waterfronts and the use of fertile soil, intense urban development has had devastating repercussions on natural processes that manage stormwater and floods. Creating parks can restore these natural floodplains and let nature run its course. Floating structures and elevated housing can limit human disruption of the natural habitat and processes that manage water.

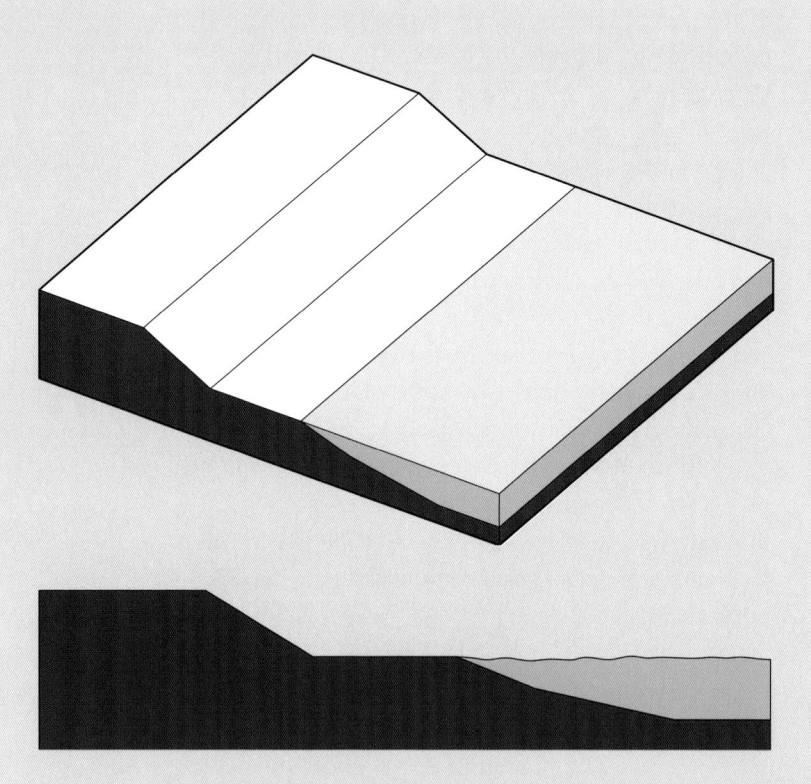

PERFORMANCE   Floodable plains are inexpensive and easy to maintain, but do require significant amounts of land, which can be difficult to find in dense urban areas. If waters continue to rise, the floodable plain may be inundated for longer periods of time, reducing human usage of the plain. Additionally, flooding may damage infrastructure in place.

PROS
- Increase groundwater recharge
- Increase the fertility of soils, leading to plant growth and healthier environments
- Provide area for recreational opportunities that could support the economic base

CONS
- Floodwaters can be a hazard for pedestrians and drivers because the depth and pace of currents can be deceptive
- Floods can damage infrastructure, for instance, by knocking down power lines, thus making the inundated plain dangerous for people to walk in, or by sweeping away cars and potentially drowning people.[1]

DESIGN GOALS
- Provide access to water
- Incorporate places for human interaction and recreation that is tolerant of floods
- Program the edge with viewing decks

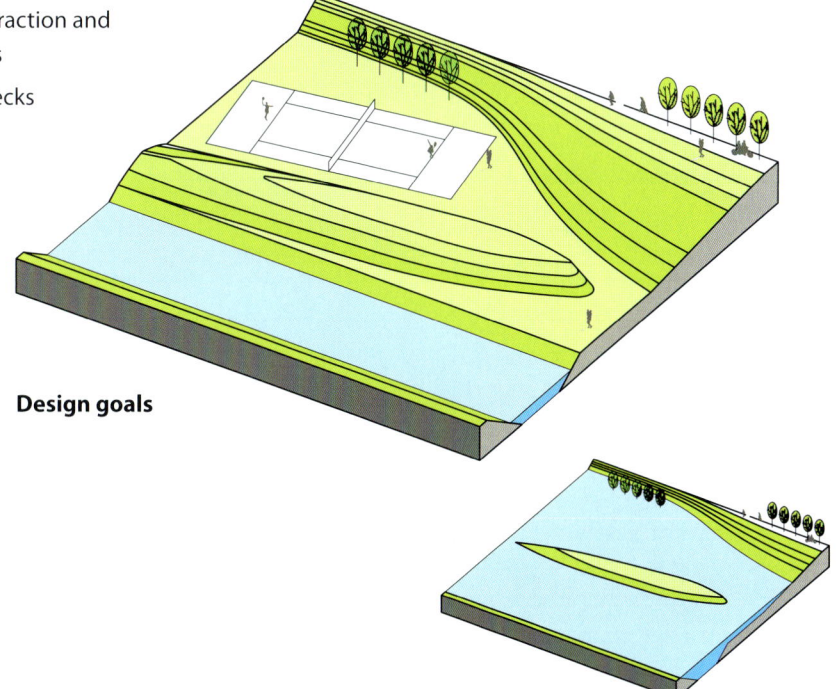

Design goals

Floodable park

Seasonal flooding in monsoon climates allows for floodable parks during the rainy season.

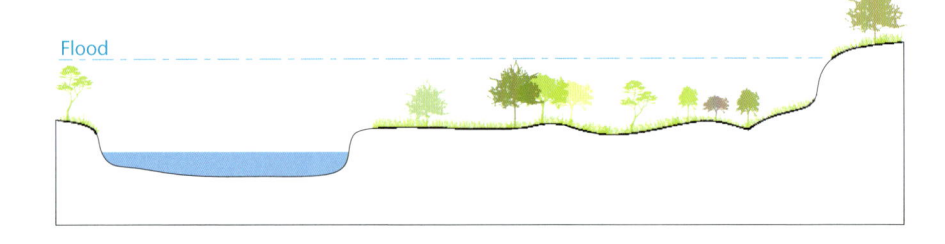

Recreation

Floodplains can be enhanced with recreation or amphitheaters to allow for viewing.

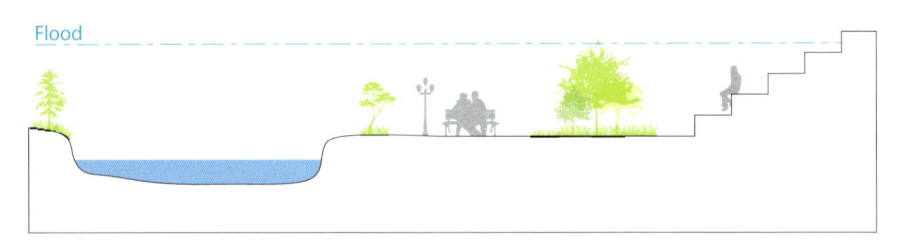

Elevated building

Sporadic structures within the floodplain can help activate the parks with uses.

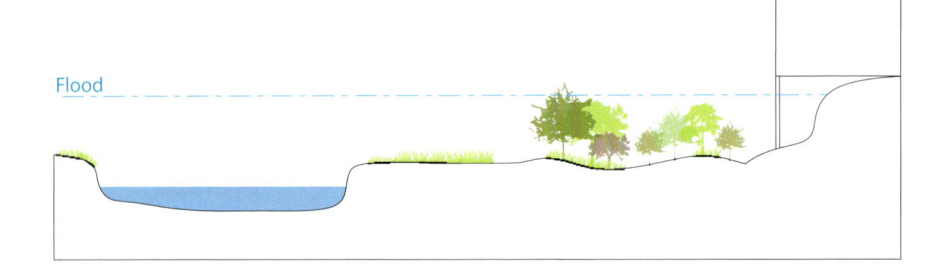

Floating pools

Multilevel water pools can accommodate various levels of flooding and enhance the experience of water.

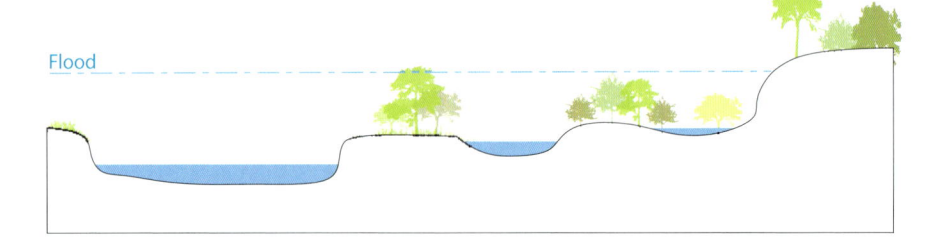

CUMBERLAND PARK, NASHVILLE, TENNESSEE

A former riverfront wasteland in Nashville, located on the Cumberland River's east bank, has been transformed into a play park for families, while helping to preserve the floodplain, improve water storage, harvest water for irrigation, remediate the brownfield, and improve biodiversity. The new 6.5-acre (2.6-ha) Cumberland Park and adjacent Bridge Building have become a vibrant recreational ground downtown. Landscape architects Hargreaves Associates also designed a playground that inspires nonscripted play and interaction with the water, including water sprays, a rain curtain, and a viewing deck extending out over the river.

Cumberland Park

An elevated, sinuous walkway brings visitors through the park, and will likely give them a dry passage during times of flooding.
(Source: Hargreaves Associates)

8.2 POLDER

Polders are low-lying strips of reclaimed land or drained marshes enclosed by levees. They are mostly associated with the Netherlands, where they have been created since the eleventh century; today, their popularity has spread outside of northern Europe. Polders include a hydraulic transport system of drains to contribute to water storage and transportation to water pumping stations, helping to keep water levels steady and store water from rain and rivers. While in the fifteenth century the Dutch polders were kept dry by iconic windmills, today electric and diesel engines help pump water. Polders can hold water temporarily, as well as divert water. In the past, they were deliberately flooded during times of war, turning the polder into a hard-to-navigate marsh for troops.

  The role of polders in managing water is under debate, however. In Bangladesh, polders had an adverse effect and likely led to the lowering of land.[2] In the Netherlands, following a new philosophy of controlled flooding instead of flood control, some polders are being "depoldered"—returned to their natural state as floodplains, such as the depoldering of Noordwaard (part of the Room for the River project, a strategy discussed in chapter 9.)

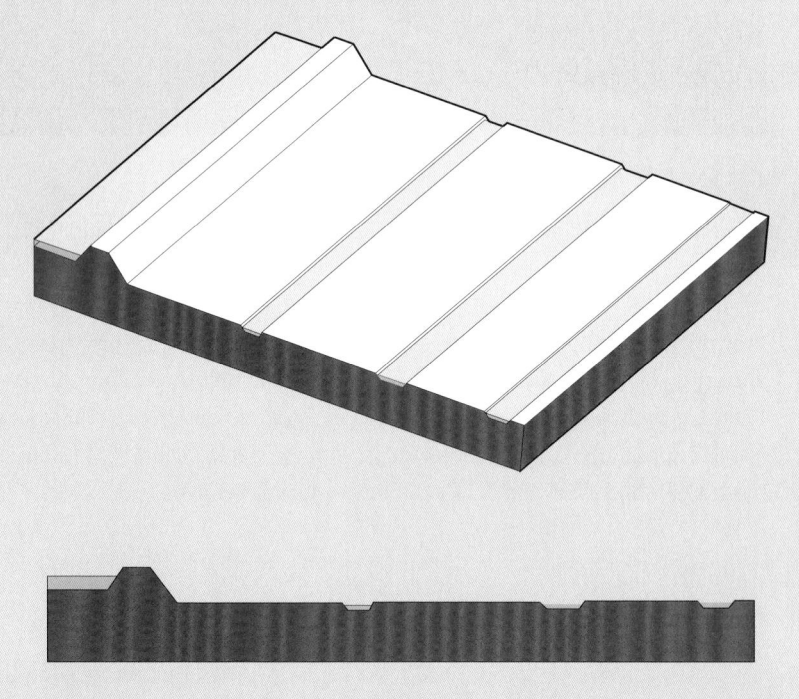

PERFORMANCE Although polders can store water, they may also help prevent storm surge if they are deployed on a large enough scale.[3] Polders could act as a first level of defense to absorb wave energy and contain floodwater. They are to be used in conjunction with other flood protection measures, such as dikes.

Polders are an attack strategy that includes land reclamation. When polders are created from reclaimed sea, the soil must be rid of salt to allow plants to grow for agriculture.

PROS

- Can help form a freshwater lake to become a valuable source of fresh water, especially during occasional summer droughts
- Can help to improve brackish marshlands
- Can be developed into freshwater fisheries

CONS

- At risk from flooding at all times

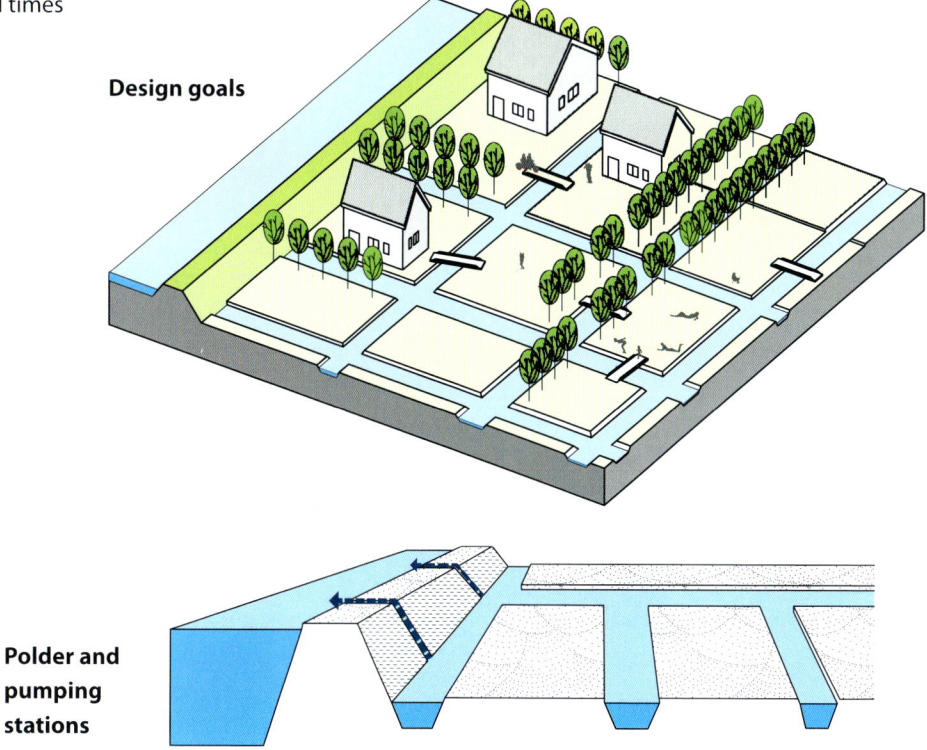

DESIGN GOALS

- Program individual patches with functions that are floodable
- Integrate isolated patches with bridges, boats, and roads
- Pumping function is critical to recovering quickly from flooding events

Design goals

Polder and pumping stations

Stormwater storage

Polders can act as a reserved space for floodwater that can be quickly recovered from any flood event. Facilitating flood and recovery has the highest priority in polders. There is a danger of pollution from agricultural fields containing fertilizer or other chemical elements.

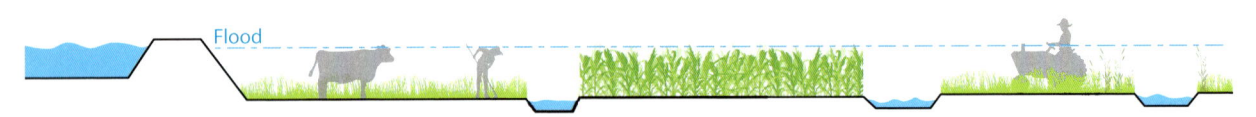

Farmland

Agricultural functions that allow quick evacuation are preferred land uses for polders.

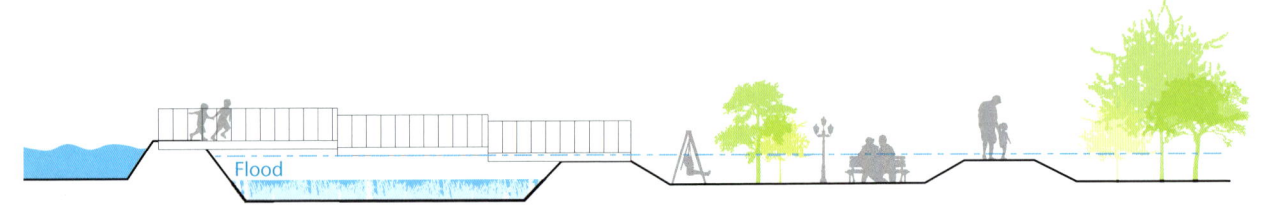

Recreational park

Polders can also function as parks for recreation when not inundated, almost as a rural equivalent to the water plaza.

POLDER LANDSCAPE, THE NETHERLANDS

The saying, "God created the world, but the Dutch created the Netherlands," refers to the country's polders. Today, about two-fifths of the Netherlands is polder land, once converted from swamp land or ocean. A lot of this land has an agricultural purpose, growing flowers, vegetables, and plants. The Dutch also use the polders for recreation, including bike rides. However, this feat of hydrological engineering, with land lying lower than sea level, has brought the nation an expensive battle with water, such as a hydraulic system that relies on continuous pumping. Recently, a new philosophy of "depoldering" is taking hold. In contrast to the engineering approach, it lowers dikes to restore floodplains and make room for riverine nature.

Polder in Zeeland, The Netherlands

Recreational bikers enjoy the view of the vast polder landscape from an elevated road in Zeeland, The Netherlands.

(Source: xpgomes, Flickr)

8.3 FLOODABLE SQUARE

Floodable squares and parks are lowered urban areas that become pools during heavy rainfall or flooding from the sea or river. They can be used for stormwater storage in inner cities, as in Benthemplein, the first full-scale water square in Rotterdam, the Netherlands. They can also function on the river shore or seaside. As water levels change, the square becomes partially or fully flooded. A water square can also function as an urban public space, because the lower retention areas can be used for sports and recreation during dry conditions.

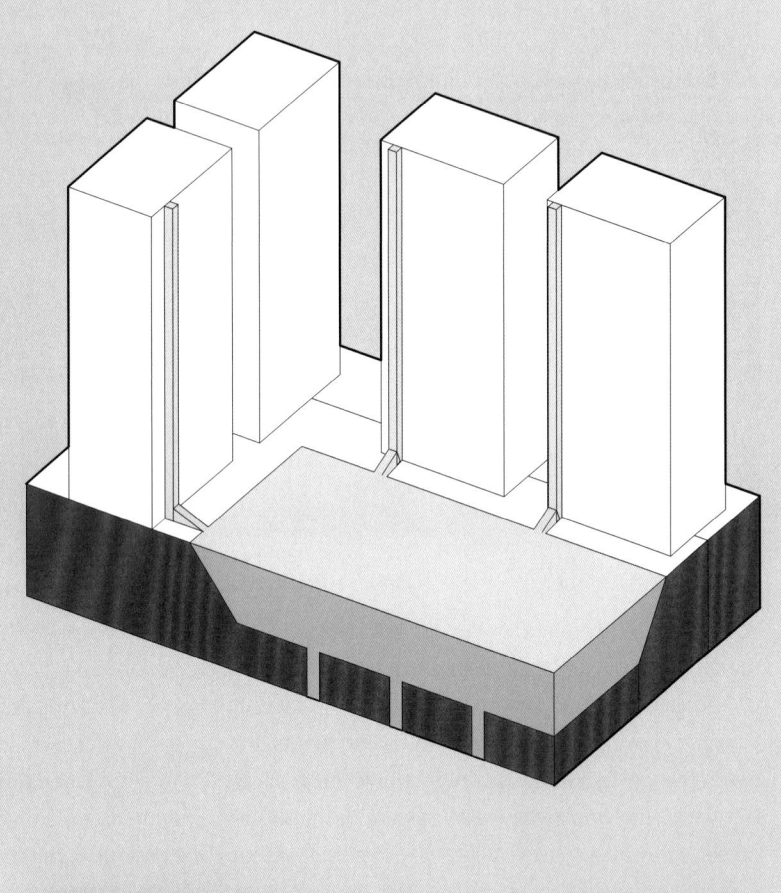

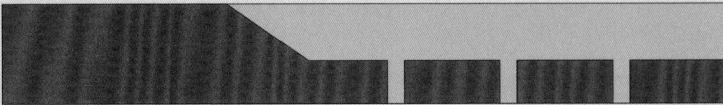

PERFORMANCE Floodable squares combine a number of stormwater management tactics to successfully provide flood protection, such as water storage basins and stormwater infiltration in the form of drainage systems and permeable pavements and soils. They can require significant upfront construction and maintenance costs, making them less common in the United States.

PROS
- Instead of hiding runoff water underground, water squares can make water storage infrastructure visible to the public and invite interaction
- Floodable squares are even suitable for densely populated areas
- They can also serve as a reservoir for rainwater from the roofs of surrounding buildings

CONS
- If not maintained, floodable squares can easily accumulate dirt, leaves, and trash
- May pose a risk to children if communities are not educated about use

DESIGN GOALS
- Program the square with recreation or sports
- Incorporate places for human interaction with engineered stormwater infrastructure, for instance, the water channels
- Provide access to water at various levels

Design goals

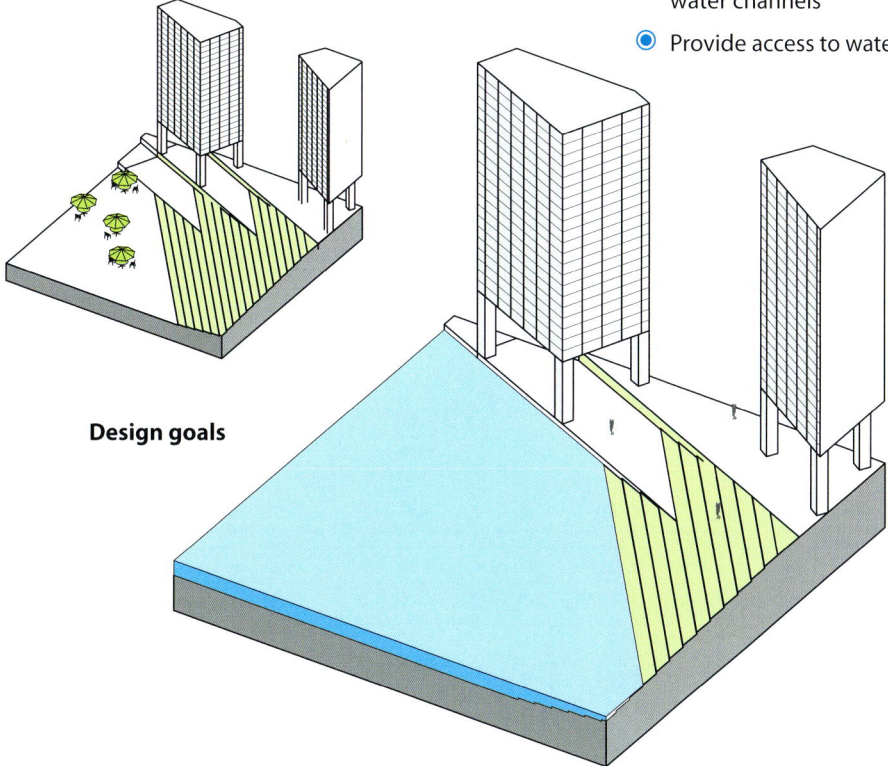

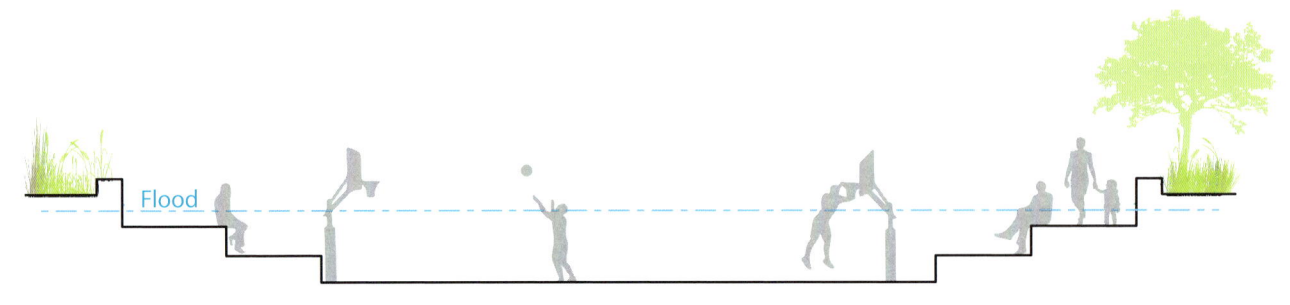

Floodable playground

Multipurpose structures that act as playgrounds or sports fields during dry periods can be flooded during storms.

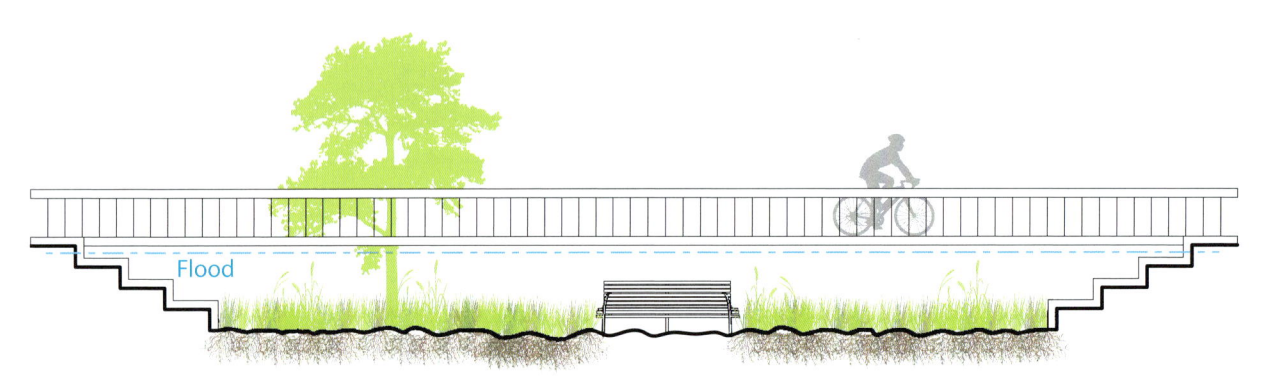

Bioretention pond

A floodable square could collect and transport water during floods and storms while promoting biodiversity.

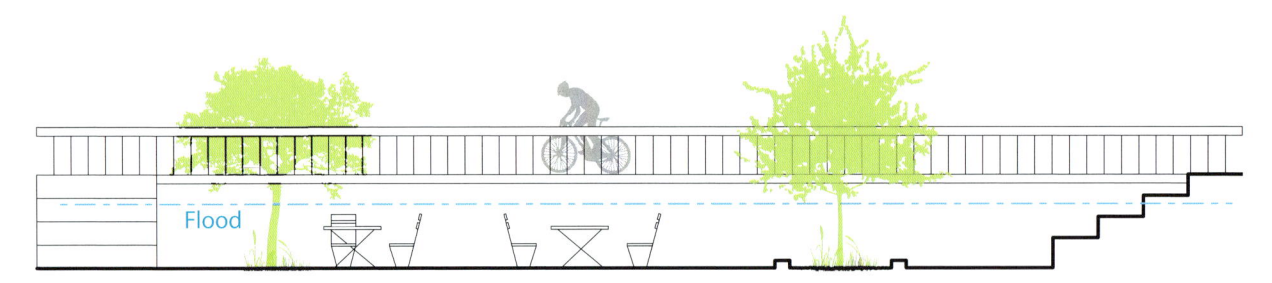

Sunken plaza

Sunken places could have seating areas that can hold water during times of flooding.

BENTHEMPLEIN WATER SQUARE, ROTTERDAM

The Benthemplein Water Square was built in 2013 in Rotterdam, the Netherlands, as a result of community design and involvement to help Rotterdam support its flood protection programs. In addition to managing stormwater and floods, the square also helps reduce heat island effects and serves as a usable and enjoyable public space. The multifunctional water square is used as a sports venue, theater, and park when not needed for water retention. During times of heavy rain and storm surge, the water square collects rain water and stores it in basins within the square. Two of the basins collect water while the third is used for excess. Stormwater then runs through steel gutters under and along the square, draining into an underground infiltration system and out of the city. The blue-shaded and curvaceous graphic pattern of the square's surface is inspired by weather charts.

Benthemplein Water Square, Rotterdam

The Water Square is a water basin, a basketball court, and an urban plaza, allowing multiple uses as storms come and go.

(Source: De Urbanisten, © Pallesh+ Azarfane)

8.4 STORMWATER INFILTRATION

Natural areas that allow stormwater infiltration, also known as "green" infrastructure, provide a form of stormwater management that allows stormwater to filter through natural and human-made structures to avoid surface flooding. In contrast to standard "gray" stormwater infrastructure, such as pipe drainage systems, stormwater infiltration treats runoff through natural systems such as rain gardens, right-of-way bioswales, and retention basins. Infiltration areas can also include human-made structures such as permeable pavements.[4]

As compared to traditional gray infrastructure that has only a single use, like a pipe transporting water, green infrastructure has multiple benefits, from improving water quality to improving quality of life. When integrated into stream networks, green infrastructure can purify water, reduce runoff, control floods, and manage stormwater. With the right vegetation, it increases biological diversity and hosts urban agriculture. As green roofs, it reduces energy consumption by cooling and insulating buildings. And with the appropriate trees, green infrastructure can help filter air pollution, absorb noise, and reduce wind loads.

PERFORMANCE Stormwater infiltration systems range in type and performance. Some stormwater infiltration systems, such as right-of-way bioswales, can be fairly inexpensive to install and can provide some local benefits. Others, such as infiltration basins, can be very expensive and require significant installation and maintenance. In cities such as New York City, the concern with drainage systems (particularly combined sewer outfalls) is their risk of backing up and causing greater local flooding. Also, if not properly maintained, they can become obstructed. Permeable streets are another infiltration option. They have large initial costs, but comparable or even cheaper maintenance costs than traditional infrastructure, particularly for cities with a combined sewer system that collects both sewage and stormwater.

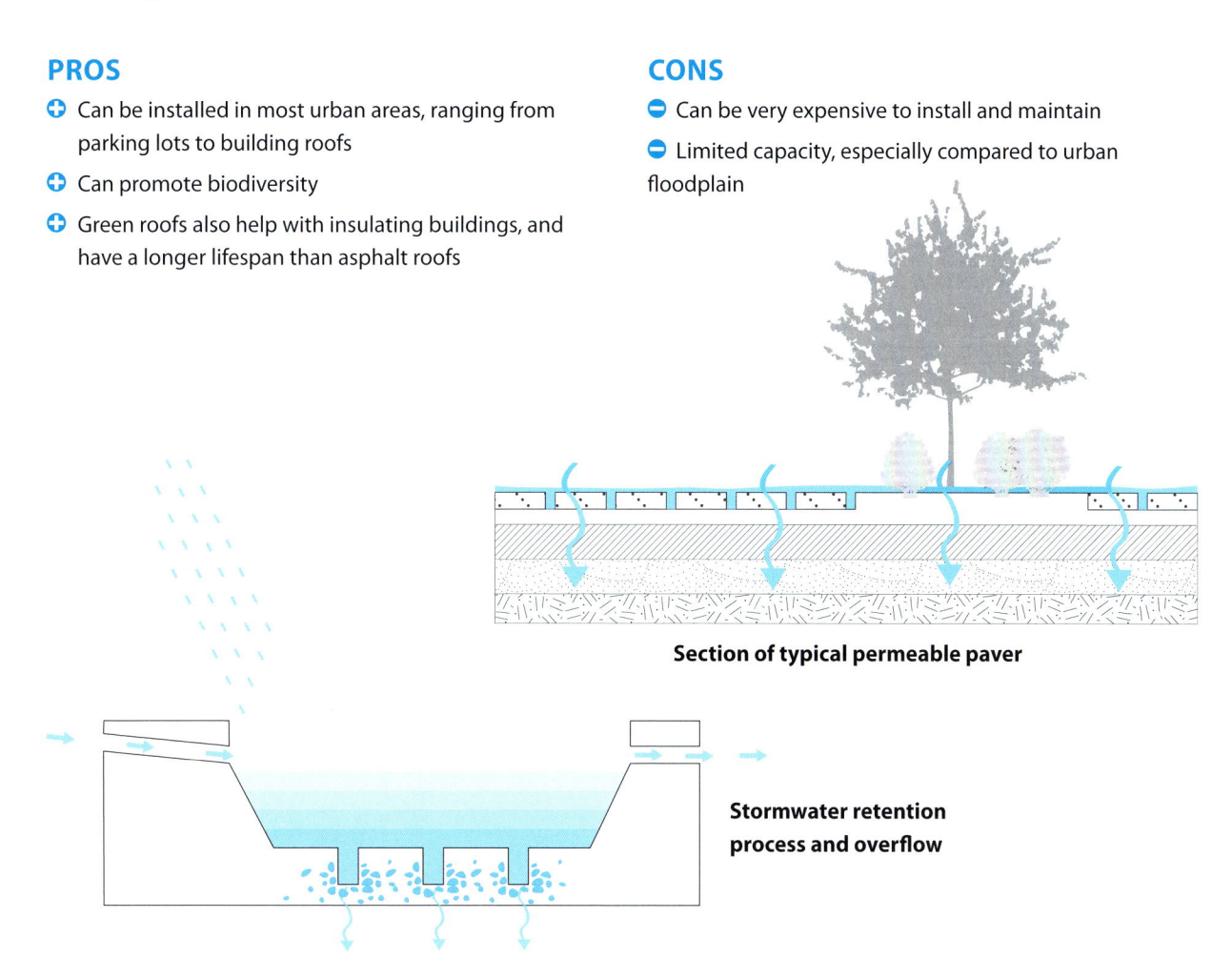

PROS

- Can be installed in most urban areas, ranging from parking lots to building roofs
- Can promote biodiversity
- Green roofs also help with insulating buildings, and have a longer lifespan than asphalt roofs

CONS

- Can be very expensive to install and maintain
- Limited capacity, especially compared to urban floodplain

Section of typical permeable paver

Stormwater retention process and overflow

Permeable parking

Permeable parking lots allow stormwater to permeate through road or lot surface and into underground water transportation networks.

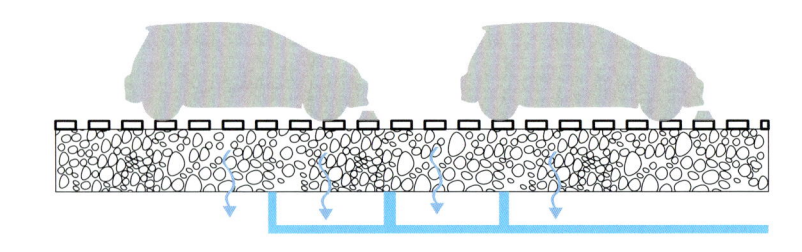

Permeable plaza

Permeable plazas allow stormwater to permeate through the plaza surface and into water transportation networks belowground.

Bioinfiltration park

A bioinfiltration park captures and diverts runoff while filtering polluted stormwater.

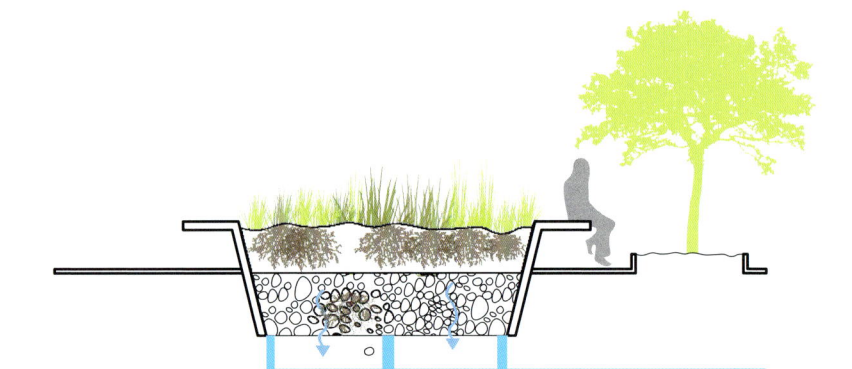

Bioinfiltration planter

In areas where infiltration of stormwater cannot be achieved, because of constrained sites next to buildings for instance, walled planters with a subsurface drain system can manage stormwater and treat water quality.

INFILTRATION BED, PHILADELPHIA, PENNSYLVANIA, USA

Philadelphia's Water Department has developed an elaborate stormwater plan—Green City, Clean Water—which includes a subsurface infiltration program. Subsurface infiltration includes stone beds with water storage and transportation pipes below. As water fills the voids among the stones, it seeps to the bottom and soaks into the soil. Infiltration projects can be located underneath pervious surfaces such as lawns, as well as impervious surfaces such as parking lots. The infiltration beds are pervious soils, usually containing vegetation atop to provide temporary stormwater storage below. In Philadelphia, the major cost components include site excavation, stone aggregate, non-woven geotextiles, pipes, and plantings.[5]

Infiltration bed

In Philadelphia, interlocking concrete permeable pavers allow water to infiltrate the ground instead of running off.

(Source: Philadelphia Water Department)

Park and retention basins

In a park/retention basin, stormwater can flow into a containment basin and be held and absorbed into the ground below the basin.

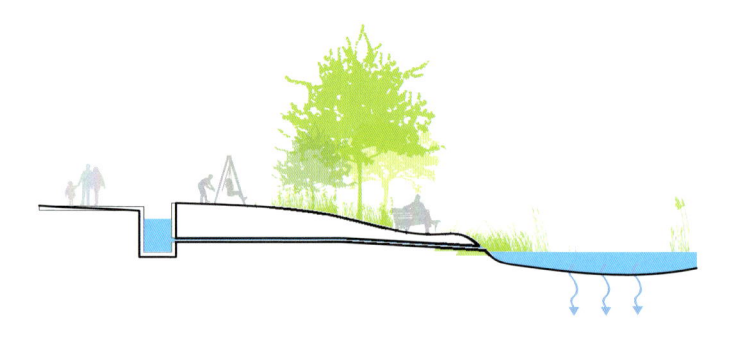

Multi retention ponds

Multiple stormwater retention ponds help increase stormwater carrying capacities and reduce risk of overflow.

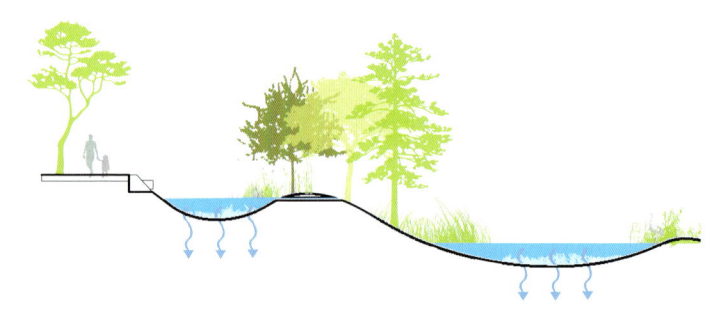

Recreation and basins

Small stormwater catchment systems at-grade feed into below-grade containment systems, allowing stormwater to be transported away.

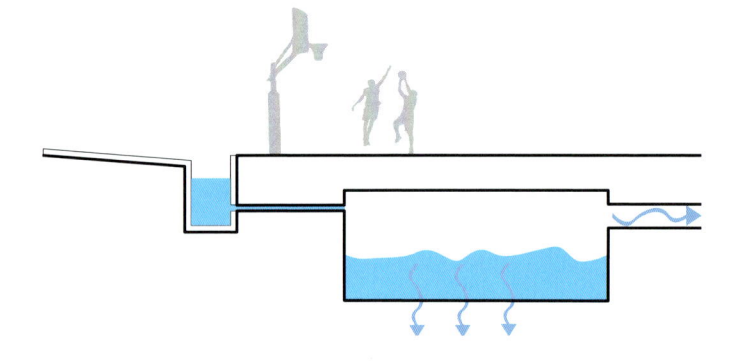

Stormwater planters

Small planters can allow for stormwater retention for personal use.

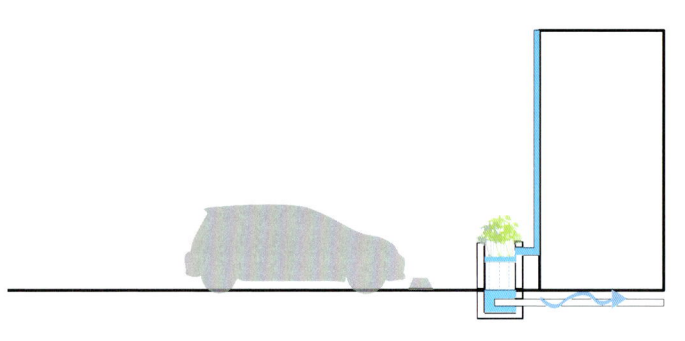

SHOEMAKER GREEN, PHILADELPHIA, PENNSYLVANIA, USA

Shoemaker Green, located on the University of Pennsylvania campus in Philadelphia, used to be a "grayfield" of aging tennis courts. Today, it is a green "front yard" to the university's historic athletics precinct, with the potential for active recreational opportunities. The green has been optimized to capture and house stormwater runoff belowground, so as not to interrupt usage above. Shoemaker Green captures 90 percent of stormwater runoff from the surrounding buildings and filters water for reuse in irrigation and in the campus heating, ventilation, and air conditioning system. Shoemaker Green incorporates native plantings and state-of-the-art stormwater management systems to reduce environmental impact as well as heat island effects.[6] The university also implemented a five-year monitoring plan, measuring parameters including stormwater quantity and quality, soil biology, and carbon sequestration.

Shoemaker Green

Shoemaker Green features a central lawn, permeable pavers, and a large rain garden.

(Source: Planphilly, Flickr)

CHAPTER 9: RETREAT STRATEGIES

▶ In contrast to "attack" and "defend" strategies that advance or protect the line of defense, retreat is an "accommodation" strategy that lets water in. The planned, "strategic" retreat solutions that are featured in this section include raising ground plains, floodproofing buildings (e.g., amphibious homes), and developing Room for the River, a Dutch program that relocated dikes to create additional space for water within the river's floodplains. But retreat can also be unplanned, as a response to a disaster, as will likely happen to the vast majority of the world's people living in flood zones if no adequate flood management strategies are put in place. Retreat avoids the costs of flood protection and is the most long-term solution, because it essentially moves people and assets away from areas of flood risk, whether uphill, above, or on water.

9.1. Raised Ground

9.2. Floodproofing

9.3. Strategic Retreat

9.1 RAISED GROUND

Raising the ground plane is a strategy that invites water to penetrate waterfront districts while elevating infrastructure such as roads to sustain human use during floods. This technique provides the opportunity for development for residential, office, hotel, retail, and transit uses. Raised ground planes can support structures and new areas for people to walk, jog, and ride their bikes. Walkways and buildings with varied elevations provide dynamic pedestrian experiences that can work well for office buildings and shopping centers.

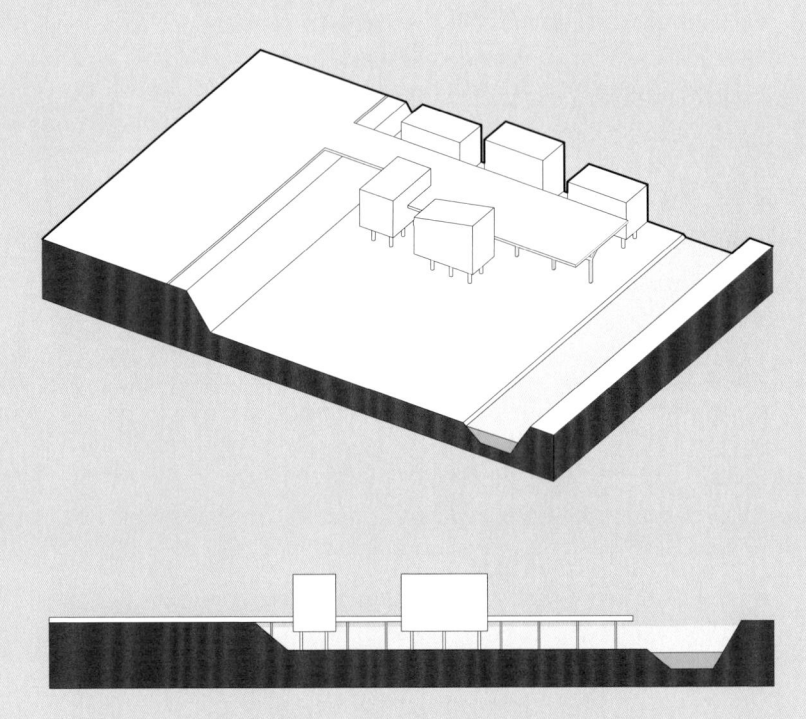

PERFORMANCE Raising the ground plane can be an effective tool for flood protection, but it's expensive and can be insecure. There must be constant reinvestment in infrastructure and buildings, including continuing to raise the structure(s) as sea levels and/or storm surges increase. Often, roads and other critical infrastructure are raised, such as in Miami Beach, with the expectation that residential buildings will follow to meet the raised ground, but some owners of existing buildings can find costs prohibitive. Raised ground plains are easiest to implement in new districts and can establish interesting views over lower-level grounds, such as in HafenCity, Hamburg, Germany.

PROS
- Increases usable area
- Can provide views
- Enhances air quality and circulation
- Better management of water table

CONS
- Conventional networks for utility distribution must be rethought
- Extremely expensive
- Difficult negotiation between public and private entities

DESIGN GOALS
- Raise usable space out of the threat zone by building ground floor on stilts, replacing ground level with other functions such as parking, or filling entire ground with structure or embankment
- Provide accesses to overcome the height difference between levels
- Activate building edge

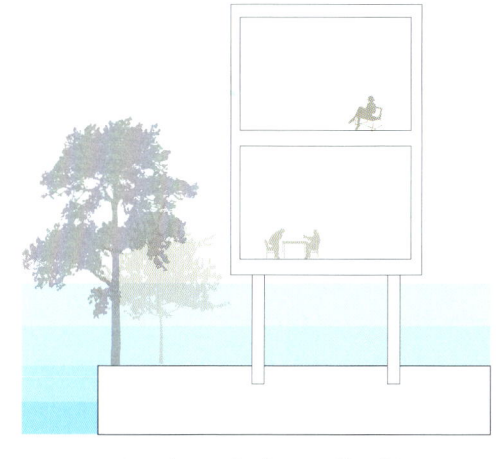

Section through elevated building

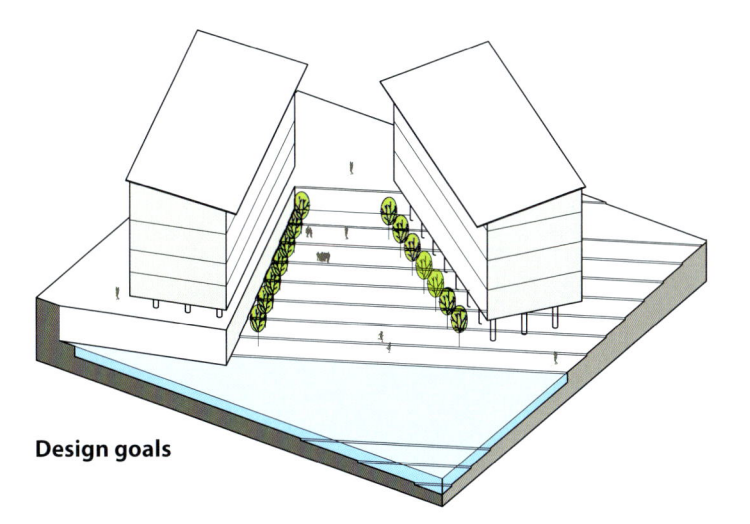

Design goals

PART II: RETREAT STRATEGIES | 123

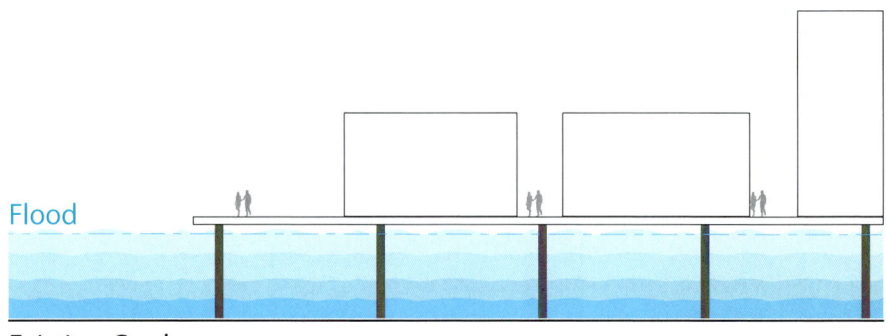

Elevated platform

The public realm is elevated on top of a new platform.

Flood

Existing Grade

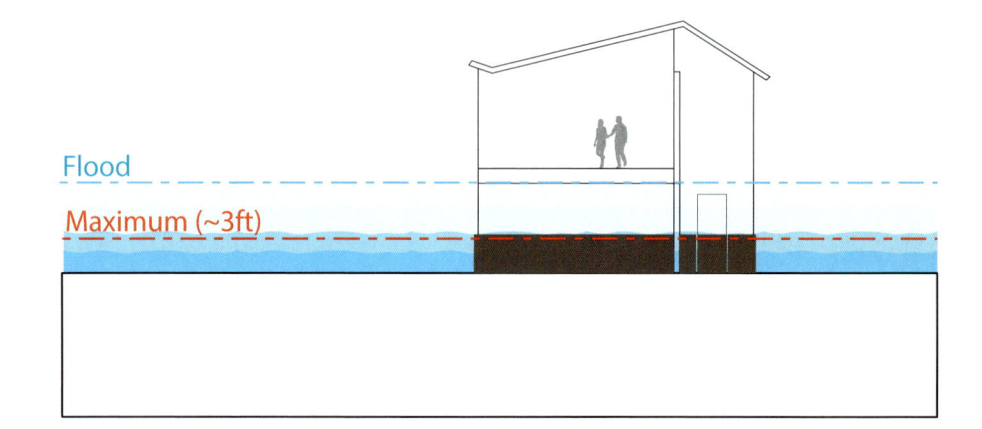

Fill-in new grade

Existing grade is elevated with earth fill from adjacent dredging or construction excavation.

Flood

Maximum (~3ft)

Strategic elevation

"Strategic" elevation of critical infrastructure and a new elevated evacuation ground network operable during disaster.

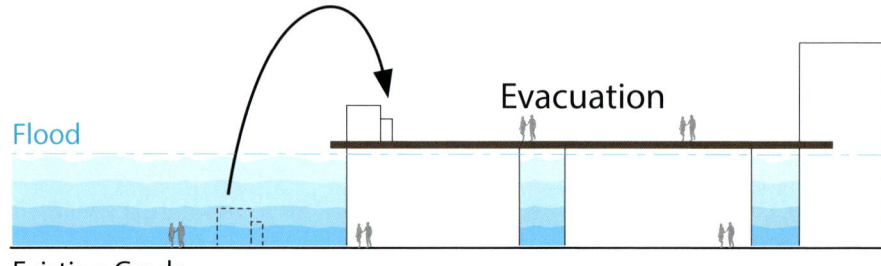

Flood

Evacuation

Existing Grade

HAFENCITY, HAMBURG, GERMANY

HafenCity in Germany, is an urban regeneration project of a former port area in Hamburg. Threatened by periodic flooding from storm surges, the project has been a showcase of sea level rise adaptation. Since its beginnings in 1997, various resiliency strategies were tested. Initially, a dike system was considered, but the costs proved prohibitive. Instead, the redevelopment authority required new roads and public spaces to be elevated over 25 feet (7.6 m) above the normal high tide, and buildings to be waterproofed up to the raised ground level.

HafenCity

Steps go down from the raised ground into a lower-level plaza that can be flooded in HafenCity, Hamburg.

(Source: Kai Bates, Flickr)

9.2 FLOODPROOFING

Floodproofing is a common technique to prevent flooding of individual structures. There are four different types of floodproofing: 1) wet floodproofing, 2) dry floodproofing, 3) elevating the structure, and 4) amphibious or floating structures.[1] Wet floodproofing allows floodwater to enter and leave a structure through designated openings and thus requires nonoccupied space. Dry floodproofing prevents water from entering a structure through watertight designs and is technically a protect strategy. This strategy allows for more usable space in a structure than wet floodproofing, but it cannot support extended periods of flooding, as leakage is bound to occur. Elevated structures and floating structures are more expensive. In elevated structures, all or the most vital building infrastructure is raised above the flood line. Floating structures rise and fall with the floodwaters. The cases in this section are examples of the two latter types of floodproofing: the elevated Miami Pérez Art Museum and an amphibious home in New Orleans.

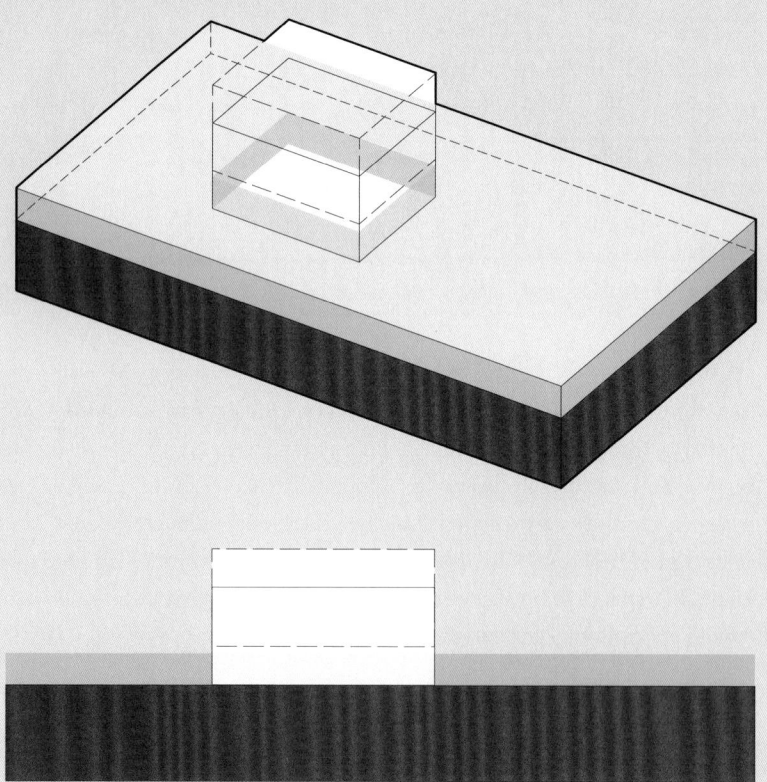

PERFORMANCE Floodproofing is the smallest scale of flood control measure, down to the building level. Many municipal governments embed floodproofing strategy in building codes. In New York City, the Department of Buildings and the Department of City Planning oversee New York's building stock by enforcing construction requirements and stewarding approval processes.

PROS

- Can be implemented more easily than other measures, including through regulation
- Does not require massive infrastructural investment from governments

CONS

- Adoption rate of floodproofing of individual buildings can be low
- Financial burden of improvement is imposed on households and private businesses
- The level of protection can be limited due to numerous entry points and underground spaces

DESIGN GOALS

- Decentralized and individualized measures to deal with flooding
- Select an appropriate waterproofing approach based on potential flooding risk
- Design building edges to be friendly to pedestrians
- Strengthen sense of community through coherent strategy

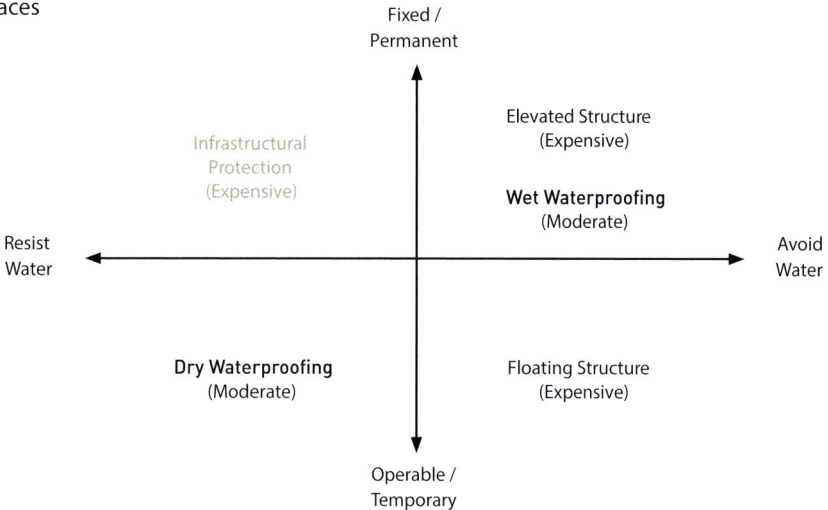

```
                    Fixed /
                    Permanent
                        ↑
                        |
                        |  Elevated Structure
          Infrastructural  (Expensive)
            Protection
            (Expensive)   Wet Waterproofing
                          (Moderate)
Resist ←────────────────┼────────────────→ Avoid
Water                   |                  Water
                        |
            Dry Waterproofing   Floating Structure
            (Moderate)          (Expensive)
                        |
                        ↓
                    Operable /
                    Temporary
```

Relative cost of waterproofing strategies

(Adapted from: Aerts et al. "Cost Estimates for Flood Resilience and Protection Strategies in New York City." *Annals of the New York Academy of Sciences* 1294, no. 1 (2013): 1–104.)

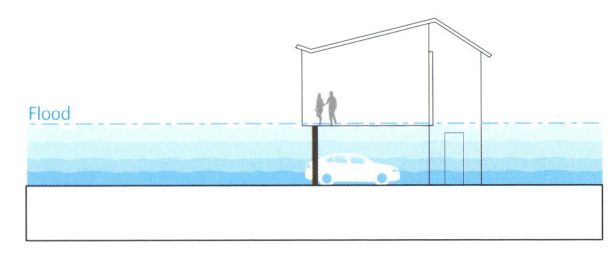

Elevated structure

Locate occupiable space above flood risk level.

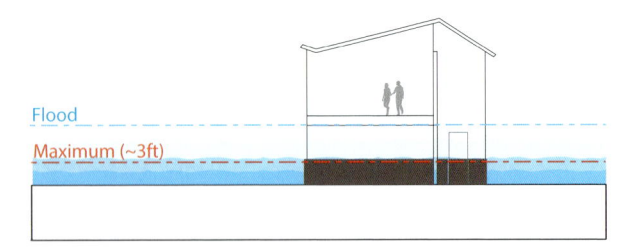

Dry floodproofing

While leaving occupiable space at ground level, minor flood threats can be mitigated with dry floodproofing, which is effective within a range of 3 feet (0.9 m) and as a temporary measure only.

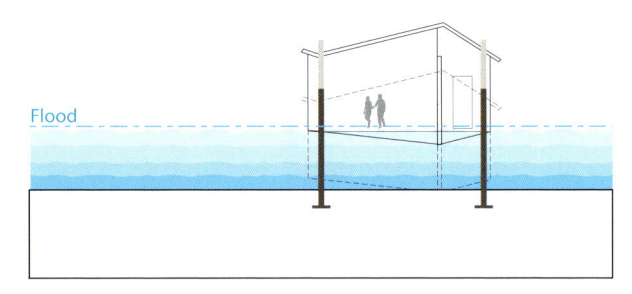

Floating structure

With simple anchoring posts, buildings can also float.

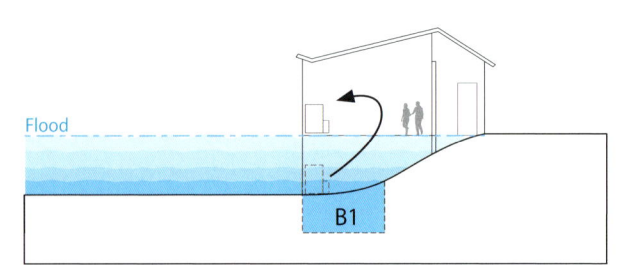

Wet floodproofing

By locating essential infrastructure out of flood risk, settlements can continue to function during floods.

MORPHOSIS'S THE FLOAT HOUSE,
NEW ORLEANS, USA

Actor Brad Pitt's Make It Right Foundation sponsored Tom Mayne from Morphosis Architects and students from University of California, Los Angeles, to build a floodproofed home in New Orleans. A new take on the "shotgun house," a narrow rectangular home typical of New Orleans, the Float House is placed on top of a raised base. The base of the house is designed to support and to provide enough buoyancy to float the entire weight of the house. The modular, prefabricated chassis is a block of expanded polystyrene foam that is coated in glass-fiber reinforced concrete. During an extreme flooding event, the base of the house acts as a raft, raising the house.

Morphosis's The Float House

Pritzker Prize–winning architect Thom Mayne and his firm, Morphosis, designed an amphibious house for New Orleans, capable of existing on water and on land.
(Source: Dmitriy Kruglyak, Flickr)

Pérez Art Museum

Pritzker Prize–winning architects Herzog & de Meuron designed the museum on an elevated platform with a grand staircase bringing guests to the entrance.

(Source: edwardhblake, Flickr)

PÉREZ ART MUSEUM, MIAMI, USA

Rising seas already increase the occurrence of recurrent, or nuisance, flooding in urbanized areas in South Florida, where cities such as Miami are not only major cruise ship hubs and seaports provisioning Florida and the southeastern United States, but also popular resort destinations for visitors from around the world. With over 22 million visitors a year, the South Florida region is the nation's second largest tourist hub. Furthermore, much of South Florida's most expensive real estate sits just above current high tide-levels on land reclaimed in the early twentieth century from coastal mangrove swamps and in shallow areas of Biscayne Bay, areas highly susceptible to recurrent flooding as sea levels rise. The National Wildlife Federation estimates that the Miami area itself has up to $3.5 trillion of assets at risk due to sea level rise by 2070,[2] the highest amount of all the world's coastal cities. Miami Beach has already invested $400 million in a system of pumps and elevated seawalls to protect its hotels and historic architecture.

Hence, Swiss architects Herzog & de Meuron designed Miami's new Pérez Art Museum with rising sea levels in mind. The museum is inspired by Stiltsville, a group of wooden houses built on stilts standing inside Biscayne Bay. A grand staircase brings visitors up the 180-foot-wide (55-m) platform that elevates the museum 18 feet above the ground, protecting it from floods. Behind the staircase and underneath the museum is a parking garage, which can be flooded. Hanging plants help to naturally cool the veranda area.

9.3 STRATEGIC RETREAT

In the future, city adaptation strategies will need to handle much more powerful tidal waves that are predicted from weather events and overall sea level rise. Rather than constantly investing in strategies that will fight storm surges, facilities could be relocated to the hinterland and secured on a flood-safe elevation. Residential, commercial, and civic buildings that are located in potential inundation areas could be strategically retreated to secure a safe future. Meanwhile, those flood-prone sites could be redeveloped, keeping flooding in mind.

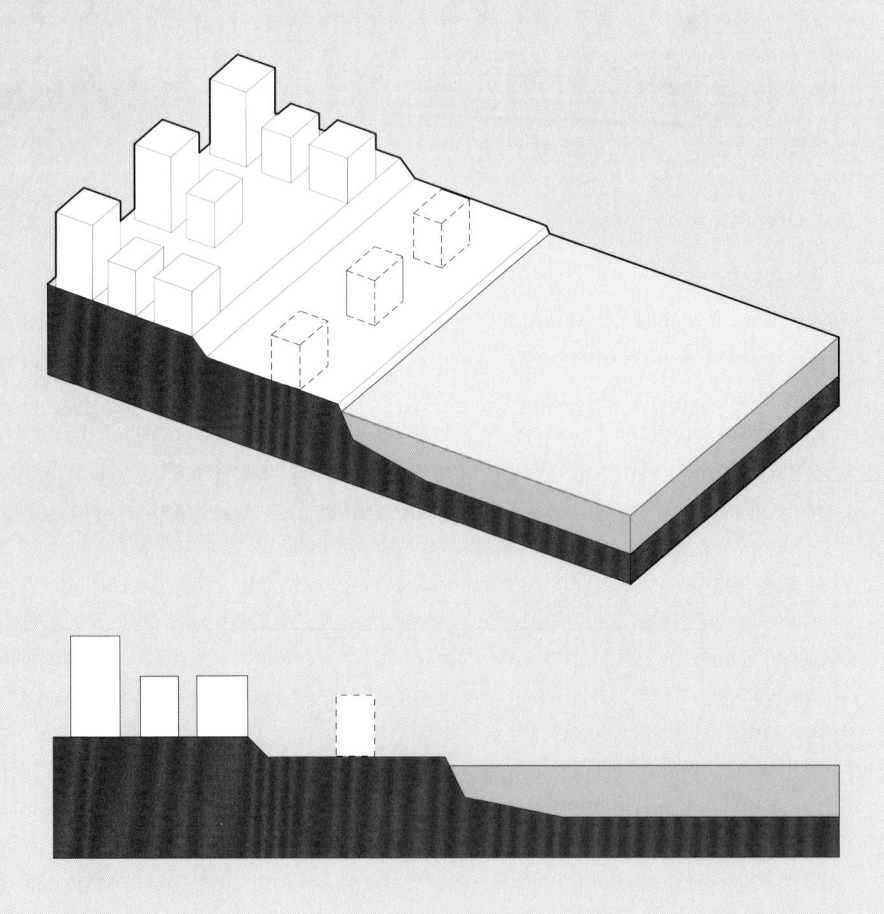

PERFORMANCE Strategic retreat, the relocation of people and infrastructure to higher grounds, is the most long-term option to address rising seas and flooding. However, strategic retreat, especially in areas with expensive real estate and/or a lack of political will, can be extremely expensive and time-consuming.

PROS
- Potential for redevelopment of underutilized inland areas
- Long-term flood protection for individuals and infrastructure once moved

CONS
- Expensive for communities/individuals as well as governments/municipalities
- Often little public/stakeholder support and buy-in
- Loss of land and infrastructure
- Politically challenging discussion

DESIGN GOALS
- Phasing is critical for retreat strategy
- Consider combining different resiliency measures, such as protect and store, to deal with coastal flooding

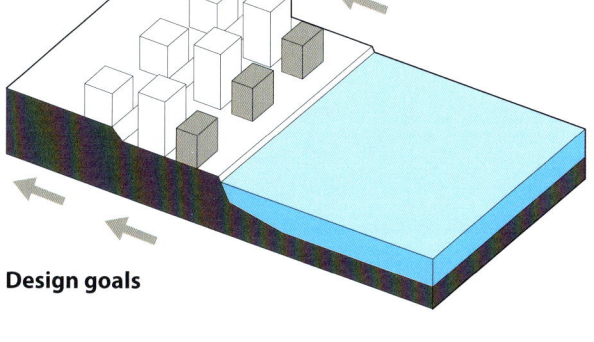

Design goals

Room for the River
By evacuating existing settlements from the floodplain, several sustainable goals are achieved: 1) new developments are protected by the new dike; 2) more areas are allowed to flood, reducing the overall risk of flooding; and 3) a compact and denser settlement helps reduce energy consumption in the long term.

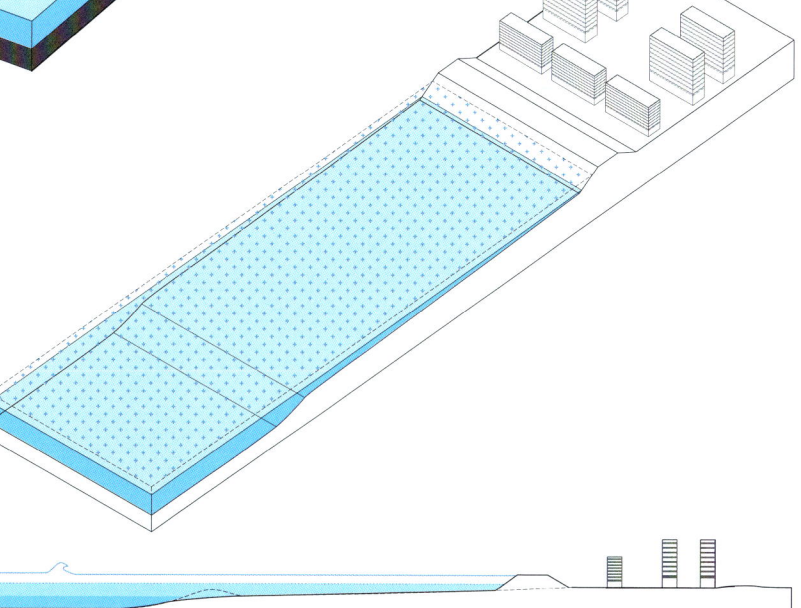

ROOM FOR THE RIVER, THE NETHERLANDS

Room for the River, or "Ruimte voor de Rivier" in Dutch, is a government-sanctioned flood protection, master landscaping, and environmental improvement project across the Netherlands's rivers, including the Rhine, the Meuse, and the Scheldt. Features of the project include installing and relocating dikes, deepening the summer bed, strengthening dikes, lowering groins, relocating polders, creating temporary water storage, and other flood protection measures. These go beyond simply ensuring future flood protection. They also improve the Dutch landscape with natural water features. The project began in 2006 and is ongoing.

The Overdiepse Polder is part of the Room for the River project. Instead of fighting to keep the water out with polders and dikes, in this new plan, planners decided to let the water in. Polders are reclaimed low-lying lands, such as marshes and floodplains, surrounded by dikes. The dike along the side of the Bergse Maas canal has been lowered, allowing water to spill into the Overdiepse Polder. This diminishes the water level in the canal by one foot (0.3 m) and reduces the risk for the upriver city of Den Bosch and its residents.

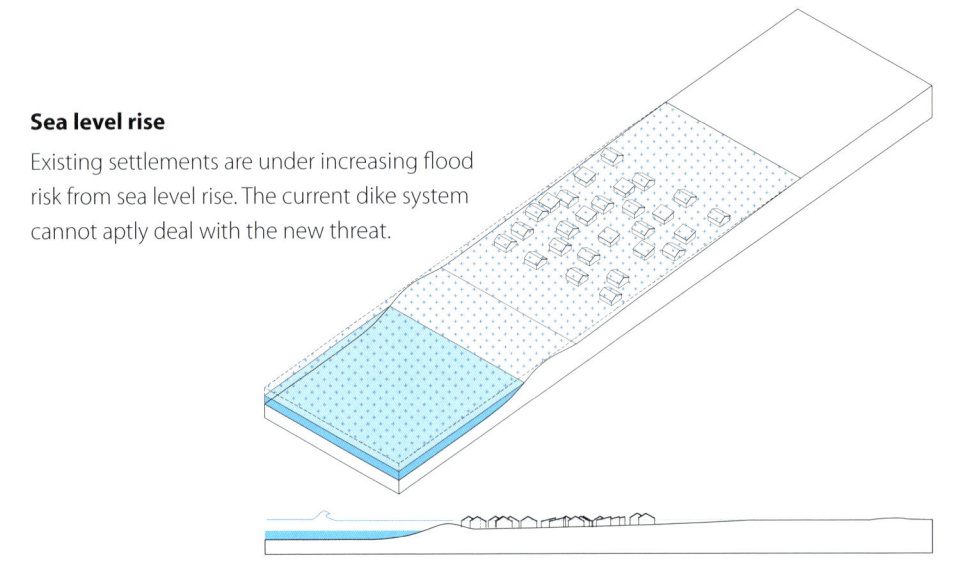

"De Oversteek" Bridge at River Park, Nijmegen
The massive bridge, part of the Room for the River project, uses local materials such as in situ concrete and brick masonry, and has a surprising human scale.
(Source: Roberto Moreno, Flickr)

Sea level rise
Existing settlements are under increasing flood risk from sea level rise. The current dike system cannot aptly deal with the new threat.

Overdiepse Polder, Room for the River

River water is spilling into the Overdiepse Polder, part of a "depoldering" strategy.
(Source: https://beeldbank.rws.nl, Rijkswaterstaat / Joop van Houdt)

Retreat

To build a stronger dike system and to acquire floodable area in front of the new dike, existing settlements are relocated behind the new dike system.

Curitiba flood control

Parque Barigui, with Curitiba in the background, contributes to flood control.

(Source: Marcos Guerra, Wikimedia Commons)

FLOOD MANAGEMENT IN THE CURITIBA METROPOLITAN AREA, BRAZIL

The Curitiba Metropolitan Area (RMC) is located in the Upper Iguacu River basin. The river has a low capacity and historically floods frequently, increasing the large natural floodplain. Much of the floodplain is unsuitable for development. In the 1980s, increased growth resulted in unauthorized development, particularly in the floodplain. Rapid growth also increased impermeable surfaces, increasing flooding in the basin by 600 percent. Poor drainage in urban areas, including poorly designed infrastructure, resulted in obstruction of natural river flow. Following large precipitation and flood events in the mid-1990s, the RMC, with help from the World Bank, developed a flood management plan. An artificial channel was dug to help increase the river capacity. This channel also acts as a border, preventing growth into the park. Some areas were conserved for wetlands, which improve water quality. Lastly, a flood warning system was put into place, allowing the RMC to prepare for large flood events.

RIO BOGOTA FLOOD CONTROL, COLOMBIA

The Rio Bogota Environmental Recuperation and Flood Control Project for Colombia aims to improve the water quality and reduce the flood risk of the Rio Bogota. At the same time, it aspires to create a multifunctional and recreational asset for the metropolitan region. Rio Bogota naturally has a wide riparian area with a large floodplain. Due to the rapid urbanization of the Bogota metropolitan region, the river has been channeled, creating flood-prone areas extending from the river into dense urban areas. The project includes relocation and resettlement for about 267 people out of flooded areas along the river. Financing for the project is spearheaded by the World Bank.

Rio Bogota flood control

Parque San José de Maryland.

(Source: Gustavo Wilches-Chaux, wilchesespecieurbana.blogspot.com/2014/01/conversaciones-con-el-rio-bogota.html)

CHAPTER 10: CONCLUSION

▶ Sea level rise, the increase of storms and peak river discharges, land subsidence in delta areas, and rapid population increases and urbanization in coastal areas will greatly grow the need for flood resilience strategies. But climate change will not only increase the number of floods, storms, and hurricanes, but also of droughts, forest fires, and heat waves. Some areas will suffer from too much water, others from too little. A 2018 study in *Nature Climate Change*[1] predicts that by 2050, thirty percent of the world's land area could face land degradation and desertification—including large areas in Europe, Asia, and Africa—making living in a desert a reality for more than 1.5 billion people. Instead of suffering the consequences then, cities can anticipate these conditions, and adapt to increase their resilience now.

The projects in this book serve as a call to action toward more resilient cities. They offer new design approaches toward flood protection, showing examples of integrated, collaborative, and adaptive solutions. They remind us that challenges related to climate change can also provide opportunities beyond reducing risk alone and create better communities, better environments, and better economies. At their core, these projects marry infrastructure with place-making. They show, as Edgar Westerhof mentioned in the foreword to this book, that water can mean collaboration, safety, and prosperity—all at once.

NOTES

CHAPTER 1

1. Stocker, T., and D. Qin, ed., *Climate Change 2013: The Physical Science Basis*, Working Group I Contribution to the Fifth Assessment Report of the Intergovernmental Panel on Climate Change. (New York: Cambridge University Press, 2014.) The amount of sea level rise varies from place to place.

2. Diaz, P., and D. Yeh, Chapter 2, "Adaptation to Climate Change for Water Utilities," in *Water Reclamation and Sustainability*. ed. S. Ahuja (Elsevier, 2014).

3. United Nations, Department of Economic and Social Affairs, Population Division, *World Urbanization Prospects: The 2014 Revision*, (ST/ESA/SER.A/366, 2015).

4. Hallegatte, S., et al. "Future Flood Losses in Major Coastal Cities." *Nature Climate Change* 3 (2013): 802–06.

5. Multihazard Mitigation Council, *Natural Hazard Mitigation Saves 2017 Interim Report: An Independent Study—Summary of Findings*. Porter, K., C. Scawthorn, N. Dash, J. Santos, and P. Schneider (Washington, D.C.: National Institute of Building Sciences, 2017).

6. www.100resilientcities.org

7. van Wesenbeeck, K. Bregje, W. de Boer, S. Narayan, W. R. L. van der Star, and M. B. de Vries. "Coastal and Riverine Ecosystems as Adaptive Flood Defenses under a Changing Climate." *Mitigation and Adaptation Strategies for Global Change* 22, no. 7 (2017): 1087–94.

8. Vergouwe, R., and H. Sarink. The National Flood Risk Analysis for the Netherlands. (Rijkswaterstaat VNK Project Office 2016).

9. Aerts, J. C., W. W. Botzen, K. Emanuel, N. Lin, H. de Moel, and E. O. Michel-Kerjan, 2014. "Evaluating Flood Resilience Strategies for Coastal Megacities." *Science* 344, no. 6183 (2014): 473–75.

10. Aerts, "Evaluating Flood Resilience Strategies for Coastal Megacities."

11. Bolstad, E. "High Ground Is Becoming Hot Property as Sea Level Rises." *Scientific American* (May 1, 2017).

12. Loughran, K. "Parks for Profit: The High Line, Growth Machines, and the Uneven Development of Urban Public Spaces." *City & Community* 13, no. 1 (2014): 49–68.

13. Field, C. B., and V. R. Barros, ed. "Regional Aspects," *Climate Change 2014: Impacts, Adaptation, and Vulnerability*. Intergovernmental Panel on Climate Change, (New York: Cambridge University Press, 2014).

14. Van Leussen, W., and K. Lulofs, 2009. "Governance of Water Resources." *Water Policy in the Netherlands: Integrated Management in a Densely Populated Delta*, ed. S. Reinhard and H. Folmer (Washington, D.C.: Resources for the Future, 2009): 171–84.

CHAPTER 2

1. Dircke, P., and A. Molenaar, "Climate Change Adaptation: Innovative Tools and Strategies in Delta City Rotterdam." *Water Practice and Technology*, 10, no. 4 (2015): 674–680.

2. City of Rotterdam, 2013. "Rotterdam Climate Change Adaptation Strategy."

CHAPTER 3

1. Aerts, J. C., W. W. Botzen, K. Emanuel, N. Lin, H. de Moel, and E. O. Michel-Kerjan. "Evaluating Flood Resilience Strategies for Coastal Megacities." *Science* 344, no. 6183 (2014): 473–75.

2. New York City Special Initiative for Rebuilding and Resiliency. *PlaNYC, A Stronger, More Resilient New York*, (The City of New York, 2013) http://www.nyc.gov/html/sirr/html/report/report.shtml

3. Aerts et al., "Evaluating Flood Resilience Strategies," 473–75.

4. *One New York: The Plan for a Strong and Just City*, (2015) http://www.nyc.gov/html/onenyc/downloads/pdf/publications/OneNYC.pdf

CHAPTER 4

1. Schwartz, J., and M. Schleifstein. "Fortified but Still in Peril, New Orleans Braces for its Future." *The New York Times*, February 24, 2018.

2. City of New Orleans, "Resilient New Orleans: Strategic Actions to Shape Our Future City," 2015, http://resilientnola.org/wp-content/uploads/2015/08/Resilient_New_Orleans_Strategy

CHAPTER 5

1. Vietnam Climate Adaptation Partnership Consortium. "Climate Adaptation Strategy Ho Chi Minh City: Moving towards the Sea with Climate Change Adaptation," 2013.

CHAPTER 6

1. Short, A. D., ed. *Handbook of Beach and Shoreface Morphodynamics* (Hoboken, NJ: Wiley, 2000), chapter 7.

2. Brown, S. A., and E. S. Clyde, *Design of Riprap Revetment*. HEC 11. (Federal Highway Administration, 1989). https://www.fhwa.dot.gov/engineering/hydraulics/pubs/hec/hec11sl.pdf

3. New York–Connecticut Sustainable Communities Consortium. *Coastal Climate Resilience: Urban Waterfront Adaptive Strategies*. (Department of City Planning, The City of New York, 2013). https://www1.nyc.gov/assets/planning/download/pdf/plans-studies/sustainable-communities/climate-resilience/urban_waterfront.pdf

4. New York-Connecticut Sustainable Communities Consortium, *Coastal Climate Resilience*, 77.

5. Sciortino, J. A., *Fishing Harbour Planning, Construction and Management*, no. 539. (Rome: Food and Agriculture Organization of the United Nations, 2010), 87.

6. New York–Connecticut Sustainable Communities Consortium, *Coastal Climate Resilience*, 92–93.

7. See more: "Learn More About the Living Breakwaters Project," New York State Governor's Office of Storm Recovery, https://stormrecovery.ny.gov/learn-more-about-living-breakwaters-project (Accessed Dec. 1, 2017).

8. New York–Connecticut Sustainable Communities Consortium, Coastal Climate Resilience, 85.

9. "Seven Wonders," American Society of Civil Engineers. https://web.archive.org/web/20100802060056/http://www.asce.org/Content.aspx?id=2147487305

10. New York City Economic Development Corporation and ARCADIS. *Southern Manhattan Coastal Protection Study: Evaluating the Feasibility of a Multi-purpose Levee* (MPL), https://www.nycedc.com/sites/default/files/filemanager/Projects/Seaport_City/Southern_Manhattan_Coastal_Protection_Study_-_Evaluating_the_Feasibility_of_a_Multi-Purpose_Levee.pdf

11. New York–Connecticut Sustainable Communities Consortium, Coastal Climate Resilience, 100.

CHAPTER 7

1. New York–Connecticut Sustainable Communities Consortium. *Coastal Climate Resilience: Urban Waterfront Adaptive Strategies*. (Department of City Planning, The City of New York, 2013), 78. https://www1.nyc.gov/assets/planning/download/pdf/plans-studies/sustainable-communities/climate-resilience/urban_waterfront.pdf

2. Hibbs, M. "A Living Shoreline Laboratory." *Coastal Review Online* (May 26, 2016). https://www.coastalreview.org/2016/05/14558/

3. New York–Connecticut Sustainable Communities Consortium, *Coastal Climate Resilience*, 62–63.

4. De Schipper, M. A., S. de Vries, G. Ruessink, R. C. de Zeeuw, J. Rutten, C. van Gelder-Maas, and M. J. Stive. "Initial Spreading of a Mega Feeder Nourishment: Observations of the Sand Engine Pilot Project." *Coastal Engineering* 111 (2016): 23–38.

5. Yeh, N., P. Yeh, and Y. H. Chang. "Artificial Floating Islands for Environmental Improvement." *Renewable and Sustainable Energy Reviews* 47 (2015): 616–622.

6. New York–Connecticut Sustainable Communities Consortium, *Coastal Climate Resilience*, 97.

CHAPTER 8

1. Deschu, N. "Living with the River: A Guide to Understanding Western Washington Rivers and Protecting Yourself from Floods." Washington State Department of Ecology. (2016).

2. Schiermeier, Q. "Holding back the Tide." *Nature* 508, no. 7495 (2014): 164.

3. New York–Connecticut Sustainable Communities Consortium. *Coastal Climate Resilience: Urban Waterfront Adaptive Strategies*. (Department of City Planning, The City of New York, 2013), 104. https://www1.nyc.gov/assets/planning/download/pdf/plans-studies/sustainable-communities/climate-resilience/urban_waterfront.pdf

4. *Minnesota Stormwater Manual*. Minnesota Pollution Control Agency. (2013).

5. Philadelphia Water Stormwater Management Guidance Manual, "Stormwater Management Practice Guidance: Subsurface Infiltration," http://www.pwdplanreview.org/manual/chapter-4/4.4-subsurface-infiltration/

6. A Vision for the Future, Shoemaker Green, University of Pennsylvania. http://www.pennconnects.upenn.edu/find_a_project/alphabetical/shoemaker_green_alpha/shoemaker_green_overview.php

CHAPTER 9

1. New York–Connecticut Sustainable Communities Consortium. *Coastal Climate Resilience: Urban Waterfront Adaptive Strategies*. (Department of City Planning, The City of New York, 2013), 39–57. https://www1.nyc.gov/assets/planning/download/pdf/plans-studies/sustainable-communities/climate-resilience/urban_waterfront.pdf

2. Anderson, L., P. Glick, S. Heyck-Williams, and J. Murphy. *Changing Tides: How Sea-Level Rise Harms Wildlife and Recreation Economies along the U.S. Eastern Seaboard*, (National Wildlife Federation, 2016) https://www.nwf.org/~/media/PDFs/Global-Warming/Reports/Changing-Tides_FINAL_LOW-RES-081516.ashx

CHAPTER 10

1. Park, C. E., S. J. Jeong, M. Joshi, T. J. Osborn, C. H. Ho, S. Piao, D. Chen, J. Liu, H. Yang, H. Park, and B. M. Kim, "Keeping Global Warming within 1.5° C Constrains Emergence of Aridification." *Nature Climate Change* (2018): 1.

INDEX

Note: Page numbers followed by the letter "i" indicate illustrations.

Aerts, Jeroen, 19
Afforestation, 11
Amphibious structures, 15, 128
Amphitheaters, 60i, 65i, 76i
Army Corps of Engineers, 14, 20, 36, 40

Backflow, 3
Beach dynamics and replenishment failure, 8–9i
Beach nourishment, 92–95
Beck, Ulrich, 21
Benches, 60i, 72i
Benthemplein Water Square, Rotterdam, 113
Big U (Lower Manhattan flood protection system), New York City, 20, 36, 37i
Biofilters, floating, 98i, 99
Bioinfiltration, 116i
Bioinfiltration parks and planters, 116i
Bioretention ponds, 112i
Bjarke Ingels Group, 20, 37i
Boston, USA, 4i
Breakwaters, as hard-protect strategy, 66–69, 68i
Brooklyn Bridge Park, New York City, 88
Building with Nature strategy, 28

China, 3, 5i
City design, in future "risk society," 21–22
Cleveleys Coastal Defense project, UK, 65
Climate gentrification, 20
Climate refugees, 3
Climbing walls, 60i
Connecting Delta Cities project, 28–29
Cumberland Park, Nashville, USA, 105
Curitiba Metropolitan Area (RMC), Brazil, flood management, 134

Dakpark, Rotterdam, the Netherlands, 80
Delfland, the Netherlands, 11, 28, 95
Delta Works, Rotterdam, 18–19, 26, 28
"De Oversteek" Bridge at River Park, Nijmegen, 132

Depoldering, 106, 109
Design charrettes, 21
Development: high-density, on dike platform, 80i; intensive, backshores vs. troughs for, 93i; National Capital Integrated Coastal Development program, Jakarta, 57; retracted, to protect dune grass, 94i
Dikes/levees: as hard-protect strategy, 74–77, 76i; inner- and outer-dike districts, Rotterdam, 31; multipurpose, as hard-protect strategy, 78–81, 79i, 80i; protective ring, Ho Chi Minh City, 48–49
Disaster prevention vs. treatment, 3
Dunes: and beach nourishment, 92–95, 93i; stabilization of, 11, 14; vegetation preservation, 94i

Ecological interventions, 13i
EcoShape, 11
Elevated parks, 20
Elevated paths and walkways, 90i, 94i
Elevated platforms, 64i, 124i
Elevated structures, 90i, 104i, 123i, 128i
Elevations: raising, 15, 36; strategic, 124i

Float House, New Orleans, 128
Floating breakwaters, 67
Floating floodwalls, 72i
Floating islands, 96–99, 97i–98i
Floating neighborhoods, 15
Floating paths, 68i, 90i
Floating pools, 104i
Floating structures, 15, 30i, 98i, 128i
Floodable parks, 104i
Floodable plains, as store strategy, 102–5
Floodable playgrounds, 112i
Floodable squares/water squares, as store strategy, 15, 110–13
Floodable trails, 64i
Flood management: comprehensive solutions, 18; in Curitiba, Brazil, 134; hard-protect strategies, overview, 14, 53; long-term impacts on communities, 8–11; nature's role in, 36; place-making and, 22; retreat strategies, overview, 15, 121; in Rio Bogota, Colombia, 135; soft-protect strategies, overview, 14–15, 87; typical approach, 3
Floodproofing, as retreat strategy, 126–29, 128i
Flood risk equation, 18–19

Floodwalls, as hard-protect strategy, 10i, 70–73, 72i
Fylde Peninsula Coastal Programme, UK, 65

Gray solutions, and ecological system disruption, 12i
Gray vs. green stormwater infrastructure, 114
Great Kills Harbor, Staten Island, 36
Green solutions, pros and cons, 13i
Grein, Austria, 73
Groundwater extraction, Jakarta, 57
Guangzhou, China, 5i
Gulf Intracoastal Waterway West Closure Complex, 40

Habitat breakwater, 68i
Habitat loss, 3
HafenCity, Hamburg, Germany, 15, 125
Hargreaves Associates, 105
Herzog & de Meuron, 129
High Line elevated park, New York City, 20
Highways, incorporating into dikes/levees, 77, 80i
Ho Chi Minh City, Vietnam: overview, 46–48; climate-proof plan for District 4, 48i; flood map, 46i; protective ring dike and dike redevelopment, 48–49; resilience components, 47i; resilience plan, 20; sea level rise threat, 5i; vision for 2100, 44–45i
Houtribdijk, the Netherlands, 11
Hunts Point, the Bronx, New York City, 36
Hurricane Harvey (2017), 1
Hurricane Katrina (2005), 20, 40
Hybrid gray + green strategies, in flood management, 14

Infiltration bed, Philadelphia, USA, 117
Infrastructure, 11, 13i, 114
Integrated solutions coordination, in Risk Society, 21
Interweaving ribbons, seawalls and, 60i
Isle de Jean Charles floating island project, Louisiana, 99

Jakarta, Indonesia, 57
Jersey Shore, after Superstorm Sandy, 1i

King tides, 1

Lake Borgne Surge Barrier, New Orleans, 40, 42i
Land degradation and desertification, climate change and, 137
Living Breakwaters project, Staten Island, New York City, 69
Living shorelines, as soft-protect strategy, 15, 88–91, 89i, 90i

Machland Dam, Grein, 73
Make It Right Foundation, 128
Marina Bay Barrage, Singapore, 85
Mayne, Thom, 15, 128
McHarg, Ian, 11
Miami, USA, 4i, 129
Mobile floodwall, Grein, 73
Morphosis Architects, 128
Motorway Dike (Afsluitdijk), the Netherlands, 77

Nashville, USA, 105
National Capital Integrated Coastal Development program, Jakarta, 57
National Flood Risk Analysis (VNK project), the Netherlands, 18
Nature, role in flood management, 11, 36
Netherlands: Motorway Dike, 77; National Flood Risk Analysis (VNK project), 18; National Water Centre, 30; Overdiepse Polder, 132, 133i; parking garage, Katwijk aan Zee, 81; polder landscape, Zeeland, 109; Room for the River, 26, 132; Sand Engine project, Delfland, 11, 28, 95; sandy foreshore, Houtribdijk, 11; surge barrier, Oosterscheldekering, 85; Zuiderzee Works, 77. *See also* Rotterdam, the Netherlands
New Orleans, USA: after Hurricane Katrina, 2i; Float House, 128; flood map, 40i; Greater New Orleans Urban Water Plan, 42–43; Lafitte Blueway, 43; Lake Borgne Surge Barrier, 40, 42i; resilience components, 41i; Resilient New Orleans plan, 20, 42; sea level rise threat, 4i; vision for 2050, 38–39i
New York City, USA: Big U, 20, 36, 37i; Brooklyn Bridge Park, 88; coastal design and governance improvement, 37; coastal protection strategies, 36i; flood map, 34i; Living Breakwaters project, Staten Island, 69; PlaNYC, 34–37; resilience components, 35i; resilience plan, politics and, 19–20; sea level rise threat, 4i; vision for 2050, 32–33i
Nijmegen, the Netherlands, 132

One Architecture, 37i
100 Resilient Cities program, Rockefeller Foundation, 19, 36
Oosterscheldekering, the Netherlands, 85
Overdiepse Polder, Room for the River, 132, 133i

Paris Agreement, 2–3
Parking garage, Katwijk aan Zee, the Netherlands, 81
Parks: along dikes/levees, 80i; bioinfiltration, 116i; Cumberland Park, Nashville, 105; Dakpark, Rotterdam, 80; on dikes/levees, 76i; elevated, 20; floodable, 104i; pedestrian, with surge barriers, 84i; recreational, and polders, 108i; and retention basins, 118i; running trail and, 60i
Parque Barigui, Curitiba, Brazil, 134i

Parque San José de Maryland, Colombia, 135
Paths, trails, and walkways: on dikes/levees, 76i; elevated, 90i, 94i; floating, 68i, 90i; floodable, 64i; parks and, 60i
Pérez Art Museum, Miami, USA, 129
Permeable pavers, 115i, 116i
Philadelphia, USA, 117, 119
Place-making, flood management and, 11, 22
Plazas, sunken, 112i
Polders, as store strategy, 106–9, 107i, 108i
Power outages, 3
Protect and reoccupy/reclaim, as hard-protect strategy, 54–57, 56i
Public spaces integration, in flood protection, 11

Raised ground plains, as retreat strategy, 123–25
Rebuild by Design competition, 20, 21, xii
Recreation, flood management and, 72i, 98i, 104i, 108i, 118i. *See also* Parks; Paths, trails, and walkways
Reefs, constructed, 68i
Relocation, in protect and reoccupy/reclaim strategy, 55i
Reoccupation, in protect and reoccupy/reclaim strategy, 56i
Resilience, defined, 11
Resilience plans, 18–20, 29, 31
Retreat strategies, overview, 15, 121
Revetments, as hard-protect strategy, 62–65
Rio Bogota Environmental Recuperation and Flood Control Project, Colombia, 135
Risk Society, anticipating uncertainty in, 21
Rockaway Peninsula, Queens, New York City, 36
Room for the River, the Netherlands, 26, 131i
Rotterdam, the Netherlands: Benthemplein Water Square, 113; Dakpark, 80; Delta Works, 18–19, 26, 28; flood map, 26i; flood standard, 19, 26; Maeslant Barrier, 28i; National Water Centre, 30; resilience components, 27i; resilience plans, 19, 29, 31; sea level rise threat, 5i; Stadshavens district, 19, 30–31; vision for 2050, 24–25i; water squares, 15

Saltwater intrusion, 3
Sand Engine project, Delfland, 11, 28, 95
Sandy foreshore, Houtribdijk, 11, 12i
SCAPE, 69
Sea level rise: estimates and effects of, 3; and increasing salinity, 49; in the Netherlands, 132; per degree Celsius of warming, 2–3; projected loss in 2050 for world cities, 6–7i; threat to world cities under medium- and high-risk scenarios, 4i, 5i; tool box of interventions, 16–17i; top 10 coastal urban areas threatened by, 6–7i

Seawalls: effects of, 10i; environmental harm caused by, 8; as hard-protect strategy, 14, 58–60, 60i; planned, Jakarta Bay, 57; in Vancouver, 61
Shanghai, China, 5i
Shoemaker Green, Philadelphia, USA, 119
Singapore, surge barrier, 85
SIRR (Special Initiative for Rebuilding and Resilience), xi–xii
Soft-protect strategies: overview, 14–15, 87; dunes and beach nourishment, 92–95; floating islands, 96–99, 97i–98i; living shorelines, 15, 88–91, 90i; polders, 106–9, 107i, 108i; stormwater infiltration, 114–17, 115i, 118i
Staircase/steps, 64i, 65i, 76i, 80i, 125
Store strategies, overview, 15, 100
Storm surge protection, New York City, 36
Stormwater infiltration, as store strategy, 114–17, 115i, 118i
Strategic retreat, as retreat strategy, 130–33
Stump Sound shoreline, North Carolina, USA, 91
Superstorm Sandy (2012), 1, 19, 34, 88
Surge barriers, as hard-protect strategy, 82–85, 83i, 84i, 85
Systems strategy of flood management, 14

"T" floodwall, 72i
Tottenville, Staten Island, 36
Transparent floodwalls, 72i

Upland-water-storage (store) strategy, overview, 15, 100
Urbanization of coastal areas, 3
Urban resilience, 11

Vancouver, Canada, 61
Viewing decks, 64i, 84i
Vulnerability reduction, and economic losses, 3

Wagga Wagga, Australia, 99
Water features, dikes as, 80i
Water management, collaborative, 21
Water plaza, 60i
Waterproofing strategies, 127i
Wollundry Lagoon, Wagga Wagga, 99

Island Press | Board of Directors

Pamela Murphy
(Chair)

Terry Gamble Boyer
(Vice Chair)
Author

Tony Everett
(Treasurer)
Founder, Hamill, Thursam
  & Everett

Deborah Wiley
(Secretary)
Chair, Wiley Foundation, Inc.

Decker Anstrom
Board of Directors,
Discovery Communications

Melissa Shackleton Dann
Managing Director,
Endurance Consulting

Margot Ernst

Alison Greenberg
Executive Director,
Georgetown Heritage

Marsha Maytum
Principal,
Leddy Maytum Stacy Architects

David Miller
President, Island Press

Georgia Nassikas
Artist

Alison Sant
Co-Founder and Partner,
Studio for Urban Projects

Ron Sims
Former Deputy Secretary,
U.S. Department of Housing
  and Urban Development

Sandra E. Taylor
CEO, Sustainable Business
  International LLC

Anthony A. Williams
CEO & Executive Director,
Federal City Council